Cambridge Elements ≡

Elements of Flexible and Large-Area Electronics
edited by
Ravinder Dahiya
University of Glasgow
Luigi Occhipinti
University of Cambridge

ORGANIC AND AMORPHOUS-METAL-OXIDE FLEXIBLE ANALOGUE ELECTRONICS

Vincenzo Pecunia
Soochow University, China

Marco Fattori
Eindhoven University of Technology

Sahel Abdinia
Eindhoven University of Technology

Henning Sirringhaus
University of Cambridge

Eugenio Cantatore
Eindhoven University of Technology

CAMBRIDGE
UNIVERSITY PRESS

CAMBRIDGE
UNIVERSITY PRESS

University Printing House, Cambridge CB2 8BS, United Kingdom

One Liberty Plaza, 20th Floor, New York, NY 10006, USA

477 Williamstown Road, Port Melbourne, VIC 3207, Australia

314–321, 3rd Floor, Plot 3, Splendor Forum, Jasola District Centre,
New Delhi – 110025, India

79 Anson Road, #06–04/06, Singapore 079906

Cambridge University Press is part of the University of Cambridge.

It furthers the University's mission by disseminating knowledge in the pursuit of
education, learning, and research at the highest international levels of excellence.

www.cambridge.org
Information on this title: www.cambridge.org/9781108458191
DOI: 10.1017/9781108559034

First published 2018

A catalogue record for this publication is available from the British Library.

ISBN 978-1-108-45819-1 Paperback
ISSN 2398-4015 (online)
ISSN 2514-3840 (print)

Additional resources for this publication at www.cambridge.org/pecunia.

Organic and Amorphous-Metal-Oxide Flexible Analogue Electronics

Vincenzo Pecunia, Marco Fattori, Sahel Abdinia, Henning Sirringhaus, Eugenio Cantatore

Abstract: *Recent years have witnessed significant research efforts in flexible organic and amorphous-metal-oxide analogue electronics, in view of its formidable potential for applications such as smart-sensor systems. This Element provides a comprehensive overview of this growing research area. After discussing the properties of organic and amorphous-metal-oxide technologies relevant to analogue circuits, this Element focuses on their application to two key circuit blocks: amplifiers and analogue-to-digital converters. The Element thus provides a fresh look at the evolution and immediate opportunities of the field, and identifies the remaining challenges for these technologies to become the platform of choice for flexible analogue electronics.*

Keywords: flexible amplifiers, flexible analogue-to-digital converters, organic TFTs, amorphous-metal-oxide TFTs, circuit integration on foil

ISSNs: 2398-4015 (online), 2514-3840 (print)

ISBNs: 9781108458191 (PB), 9781108559034 (OC)

V. Pecunia acknowledges financial support from the National Natural Science Foundation of China (61750110517), the Jiangsu Province Natural Science Foundation (SBK2017041510), the Priority Academic Program Development of Jiangsu Higher Education Institutions (PAPD) and the Collaborative Innovation Center of Suzhou Nano Science and Technology (NANO-CIC). M. Fattori and E. Cantatore would like to acknowledge the financial support of the European

1 Introduction

Recent years have witnessed significant research efforts in flexible organic and amorphous-metal-oxide analogue electronics. For a long while, academic and industrial actors focused primarily on the application of organic and amorphous-metal-oxide semiconductors to digital circuits, considering analogue circuits too challenging to implement. This was up until the concepts of smart-sensor systems and the Internet of Things gained momentum, leading to the realisation of the formidable potential of organic and/or amorphous-metal-oxide analogue electronics. While a couple of demonstrations of organic differential amplifiers were reported prior to 2010 [1], [2], the first systematic investigation of the scope of organic analogue electronics can be identified in the seminal work of Marien et al., which examined in detail organic unipolar amplifiers [3], [4] and reported the first organic delta-sigma analogue-to-digital converter (ADC) [5]. Amorphous-metal-oxide analogue electronics started off only a couple of years after (the first differential amplifier was reported by Tai et al. in 2012 [6]), partly due to the later development of the amorphous-metal-oxide-semiconductor technology.

These early works on organic and amorphous-metal-oxide analogue electronics evidenced the challenges of this area, and prompted the search for circuit-design, process, device and material strategies for higher circuit performance. Subsequent efforts have been rewarded with a number of breakthroughs. For instance, solution-processed organic and hybrid organic/amorphous-metal-oxide complementary differential amplifiers have now reached voltage gain figures above 40 dB [7], and in some cases even greater than 60 dB [8], and are capable of working

Commission for the projects ATLASS (Horizon 2020, Nanotechnologies, Advanced Material and Production theme, contract no. 636130). H. Sirringhaus acknowledges financial support from the Engineering and Physical Sciences Research Council (EPSRC) through the Centre for Innovative Manufacturing in Large Area Electronics (CIMLAE, program grant EP/K03099X/1) and the project Integration of Printed Electronics with Silicon for Smart sensor systems (iPESS).

down to a battery-compatible 5 V power supply [8]. Moreover, ADCs on foil have recently reached an effective resolution of 8 bits [9].

This Element arises from the need for a comprehensive picture of this growing research area, inclusive of the most recent breakthroughs and emerging directions. Prior to 2013, when the volume *Analog Organic Electronics* by Marien et al. was published [10], flexible analogue electronics had been explored almost exclusively in organic technologies of the unipolar kind. While specific aspects on subsequent developments were captured in a few later works (e.g., by Raiteri et al. [11], Abdinia et al. [12], and Sun et al. [13]), recent years have been particularly eventful, especially in the field of amorphous-metal-oxide and complementary organic/hybrid analogue electronics. This demands a fresh look at the area and a reassessment of its potential, both of which this Element pursues. In view of the multidisciplinarity of flexible analogue electronics, the authors recognise that this research domain is not only of interest to the specialists of analogue electronics and organic/amorphous-metal-oxide circuits, but also to the physicists, chemists and material scientists who, in fact, have contributed much to its development. Therefore, this work pursues both the details and their context, and strives to provide accessibility to non-specialists by also covering the more fundamental aspects of the field.

Before delving into the heart of the matter, it is worthwhile to expand on the motivations underlying flexible organic and amorphous-metal-oxide analogue electronics. After all, conventional analogue circuits, as produced via silicon technology, are commercially available off the shelf, and can meet an extremely wide range of functional demands with outstanding performance. More than that, digital electronics has long been dominant in flexible electronics, hence it is important to clarify for which applications it is useful to pursue organic and/or amorphous-metal-oxide analogue circuits. This discussion is the focus of the rest of this introductory chapter, which aids the reader to put in context this Element as a whole.

1.1 Analogue Electronics: An Interface to the Physical World

While the modern era has been dubbed the Digital Age due to the impact of digital electronics on our way of life, it is self-evident that the physical world is fundamentally analogue. The temperature and humidity in our homes, the oxygen content of our blood, the electrical pulses from our cells, for instance, all vary continuously within a given range. The interaction of our digital devices with the physical world is, therefore, mediated by analogue circuitry, whose role is to condition the informative content of the signals of interest so as to allow their reliable subsequent processing.

When dealing with quantities from the physical world, as transduced by sensors, electronics inevitably has to deal with the problem of amplification. This involves boosting the amplitude of a signal in order to prevent its corruption by electronic noise and interference. This function is best carried out in the immediate vicinity of the signal source (i.e., a sensor) so as to minimise the amount of interference that can be coupled, and to decrease the impact of the noise that is added by the signal conditioning chain.

Suitably amplified signals have to be converted into the digital domain prior to carrying out computation and further processing. This underlies the need for analogue-to-digital conversion. ADCs generate a digital representation of analogue signals exploiting together suitable analogue and digital circuitry. The analogue part of an ADC typically determines its performance in terms of accuracy, speed and energy efficiency.

In summary, amplification and analogue-to-digital conversion are two key functions to interface electronic devices with the physical world. Indeed, they are ubiquitous in sensing applications in which the acquisition of the transduced signals is to be combined with digital processing.

1.2 Sensorisation and the Need for Mechanically Flexible Analogue Electronics

Conventional silicon-based analogue electronics has been typically allocated to silicon chips. Its development over the years has gone through the same miniaturisation trend (down to the ten nanometres of the recent technology node) that has characterised digital integrated silicon electronics.

While silicon-based analogue electronics has been dominant thus far, recent developments in the electronic industry and market have highlighted the need for alternative platforms for analogue electronics in specific applications. In particular, this concerns the demand for *sensorisation*, namely the addition of sensing capability to the objects and environments of our daily life. This capability has to cope with the multitude of quantities from the physical world to be captured and processed at once, and the variety of locations of the sensing elements involved. For the sake of illustration, a vision has emerged for the realisation of smart homes, which would have to be equipped with the ability to interact with all appliances in it and to provide for the residents' safety, comfort and entertainment. Situations such as this clearly surpass the conventional paradigm of star-connected sensor systems, which come with a centralised chip carrying out all the electronic functions, while sensors merely convert the physical quantities of interest into electric signals. This approach is obviously not viable when the number of sensors to be interconnected becomes large, as the system would be burdened, for instance, by bulky wiring, and by substantial interference along the path from the sensors to the electronics.

Sensorisation thus leads to a new paradigm for analogue electronics, which generally goes by the name of smart-sensor systems. As the word suggests, one such system would integrate a manifold of sensors, and each of them would have to come with electronic functionality (hence their attribute *smart*). In particular, besides the transduction element (e.g., a photodiode for an optical smart sensor), a smart sensor would have to include a dedicated analogue interface (e.g., a preamplifier), an ADC, data transmission capability and a power supply [14]. This would enable a modular sensing system that eliminates the complexity and unfeasibility of a star-connected scheme. Finally, smart-sensor systems are an integral part of a broader technological revolution currently under way, that of the Internet of Things. According to this vision, smart sensors and distributed computing would all be interconnected seamlessly through an Internet-like network [15], [16]. Hence, the information acquired from smart sensors disseminated in the objects and environments of our daily life would be available to us in real time, allowing us to make more informed decisions.

Smart-sensor systems require the analogue electronics embedded in them to be deployed in unique form factors and environments – for instance, on clothes, on packaging, on our skin – and, at the same time, to be produced through a cost-effective technology. Only an analogue electronics technology on foil would be able to address these demands. Realising smart sensors on plastic foil would ensure their flexibility, conformability, light weight and robustness, all properties that are needed to disseminate sensors unobtrusively in the objects and environments of our daily life. Furthermore, flexible electronics naturally lends itself to high-throughput manufacturing, such as roll-to-roll fabrication, thus potentially enabling a cost-effective deployment of smart sensors. It is thus apparent that the success of sensorisation and smart-sensor systems can be enhanced and extended by the availability of technologies for analogue circuit fabrication on lightweight and mechanically flexible substrates.

It is noteworthy that the push for sensorisation has driven a large research effort in mechanically flexible sensors, which points to the attractive opportunity of integrating sensors and analogue circuity on the same flexible substrate. Indeed, over the past couple of decades, a large number of reports have appeared on flexible sensors capable of responding to a wide range of stimuli, for instance: light [17]–[19], humidity [20]–[22], temperature [23]–[26], pressure [25], [27]–[29], gas/vapour concentration [30]–[33]

a)

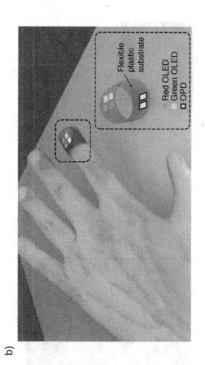

Resistance /MΩ

Temperature /°C

b)

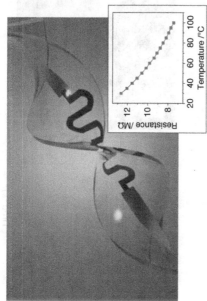

Flexible plastic substrate

Red OLED
Green OLED
□ OPD

c)

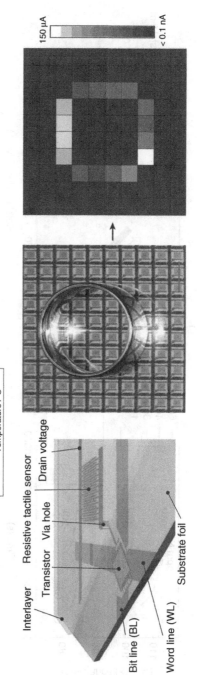

Interlayer   Resistive tactile sensor

Transistor   Via hole   Drain voltage

Bit line (BL)

Word line (WL)

Substrate foil

150 µA

< 0.1 nA

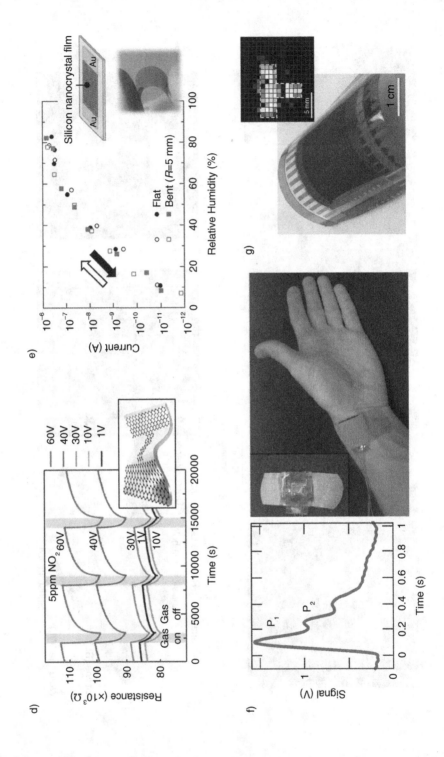

d)

Resistance ($\times 10^3\ \Omega$)

5ppm NO$_2$ 60V

40V

30V 1V
60V 10V

Gas Gas
on off

60V
40V
30V
10V
1V

Time (s)

e)

Current (A)

Silicon nanocrystal film

Au

Au

● Flat
■ Bent (R=5 mm)

Relative Humidity (%)

f)

Signal (V)

P$_1$

P$_2$

Time (s)

g)

5 mm

1 cm

Caption for Figure 1.1 (cont.)

Figure 1.1 Examples of sensors on foil. a) Graphene thermistor on stretchable foil. The plot in the inset shows the changes in resistance with temperature. Adapted with permission from [26]. Copyright 2015 American Chemical Society. b) Pulse oximetry sensor based on organic light-emitting devices and organic photodetectors. The devices are integrated onto a flexible and lightweight finger band. Reprinted with permission from Macmillan Publishers Ltd: Nature Communications ([37]). Copyright (2014). c) Organic-based active-matrix tactile sensing foil. (from left to right) Pixel structure, comprising a resistive tactile sensor and a switching transistor; photograph of a metal ring placed on the sensing foil; corresponding pixel current over the sensing foil. Adapted with permission from Macmillan Publishers Ltd: Nature ([27]). Copyright (2013). d) Flexible graphene gas sensor. The sensor is realised on plastic foil, and both its electrodes and sensing area are made of graphene (see inset). The plot depicts the sensor resistance over time at variable applied voltages in response to consecutive pulses of NO_2 gas. Adapted with permission from [33]. Copyright 2015 American Chemical Society. e) Flexible nanocrystal-based humidity sensor. The plot depicts the sensor current at variable humidity levels. The insets show a schematic of the sensor architecture and a photograph highlighting its flexibility. Adapted with permission from [22]. Copyright 2017 American Chemical Society. f) Pressure-sensing foil based on polymer transistors for the measurement of pulse waves of the radial artery. (from left to right) Signal recorded from a volunteer's wrist; photograph of flexible sensor mounted onto volunteer's wrist. Adapted with permission from Macmillan Publishers Ltd: Nature Communications ([28]). Copyright (2013). g) Flexible imager for the detection of visible light/X rays. The imager comprises organic photodetectors and carbon nanotube transistors. The inset shows the photocurrent response across the imager under illumination through a T-shaped shadow mask (see dashed line). Adapted with permission from [38]. Copyright 2013 American Chemical Society.

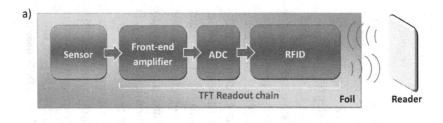

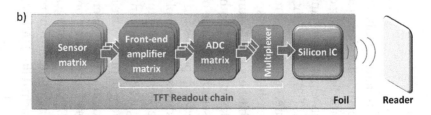

Figure 1.2 Two possible implementations of smart-sensor systems on foil. a) Sensor-augmented flexible RFID tag. b) Hybrid integration on foil of a sensor matrix readout chain and a silicon IC.

and biosignals or bioanalyte concentration [34]–[36]. A few impressive demonstrations of flexible sensors are presented in Figure 1.1, which provides an immediate visual indication of the formidable potential of flexible smart-sensor systems for manifold applications.

The block diagrams of two possible implementations of flexible smart-sensor systems are depicted in Figure 1.2. Figure 1.2a shows a sensor-augmented radio-frequency identification (RFID) tag in which all circuits are implemented using thin-film transistors (TFTs) on foil. The sensor translates an environmental parameter such as temperature into an analogue signal, which is then boosted through an amplifier and converted to digital form using an ADC. The resulting digital code is sent through a radio link to a reader, which powers the sensor system through the radio link.

The system in Figure 1.2b includes hybrid integration between electronics on foil and a silicon integrated circuit (IC). The sensor matrix measures a parameter such as pressure on multiple points on a surface. After analogue signal conditioning performed by amplifier and ADC matrices, the digital codes are multiplexed to a silicon IC, which performs further complex data analysis and transfers data through a radio link. It should be noted that alternative arrangements can be considered, where, for instance, multiplexing is applied before the ADCs. Thanks to the reduced number of interconnects, the silicon IC mounted on the edge of the foil can be small and thus inexpensive, since the whole data conversion matrix is still implemented on foil. Additionally, a sufficiently small IC (footprint of less than, say, 1 mm^2) would not compromise the overall flexibility of the smart sensor, as a minimum bending radius comparable to the size of the silicon chip would still be attainable. Finally, this implementation would also require an on-foil battery to power the whole system.

1.3 Motivation for Organic and Amorphous-Metal-Oxide Analogue Electronics

Organic and amorphous-metal-oxide semiconductors are two classes of materials that have been central to the recent developments in flexible analogue electronics. A feature they share is low-temperature processing, which makes them compatible with low-cost plastic substrates. Moreover, in recent years both types of semiconductor families have reached a performance level superior to amorphous silicon, which has long been the technology benchmark for flexible electronics. Recently synthesised organic semiconductors, both conjugated polymers and small molecules, exhibit typical p-channel mobility in the $1 - 10 \, cm^2 \, V^{-1} \, s^{-1}$ range. On the other hand, amorphous-metal-oxide semiconductors are excellent n-channel materials, with typical mobility values in the $1 - 50 \, cm^2 \, V^{-1} \, s^{-1}$ range.

Both organic and amorphous-metal-oxide semiconductors can be processed from solution, e.g., via spin coating and printing. Solution processibility is attractive in view of customisation and versatility. A roll of circuits mass-produced according to a specific template could be post-processed (for instance, via inkjet printing) to include customised features. Moreover, it can be envisaged that the value-added maker, freed from the complexities of large-volume manufacturing, could tailor mass-produced circuits to the desired application by using a compact inkjet printer. Finally, solution processing enables the facile deposition of multiple active materials on the same substrate, which is required, for instance, for high-performance complementary circuitry (comprising both n- and p-channel semiconductors) and for the integration of electronics with functional materials for sensing.

Alternatively, amorphous-metal-oxide and organic semiconductors can be deposited by vacuum-based methods. These generally lead to smaller device variability and greater control of circuit performance. Applications that have more stringent requirements on device uniformity than cost, or that do not demand the integration of multiple active materials, may therefore rightly consider the adoption of vacuum-deposited organic and/or amorphous-metal-oxide semiconductors.

Regardless of the deposition methods, both organic and amorphous-metal-oxide semiconductors offer outstanding mechanical flexibility, reliably down to a submillimetre bending radius.[1] This ensures full compliance with the requirements of conformability relevant to smart-sensor systems. In fact, this outstanding flexibility level is unique to organic and amorphous-metal-oxide semiconductors, and thus rightly qualifies these technologies as ideal for flexible electronics. Therefore, it should come as no surprise that the phrases *flexible electronics* and *organic/amorphous-metal-oxide electronics* are often used interchangeably. Unless noted otherwise, we conform to this practice in the rest of this Element.

---

[1] The most impressive demonstrations of flexibility and conformability have been achieved with circuits realised on plastic substrates down to a thickness of 1 - 3 μm [8], [91], [152]. The extreme lightness of these circuits (≈ 1 g m^{-2}) indicates that this approach represents the ideal platform for unobtrusive smart sensors.

In summary, organic and amorphous-metal-oxide semiconductors are fully compatible with circuit fabrication on foil, and enable the adoption of high-throughput and potentially low-cost manufacturing methods. All these elements point to their formidable potential for mechanically flexible analogue electronics that can meet the requirements for smart-sensor systems.

1.4 Structure of This Element

Having clarified the motivation for organic and/or amorphous-metal-oxide analogue electronics, this Element now sets out to examine the fundamentals, developments and emerging trends of this research area. At first, the properties of these two semiconductors technologies (at the material, device and processing levels) are discussed in Chapter 2, as relevant to their application to flexible analogue circuits. Subsequently, this Element covers the two prototypical circuit blocks of organic and amorphous-metal-oxide analogue electronics: amplifiers (Chapter 3) and ADCs (Chapter 4). On one hand, the extensive treatment of these two key circuit blocks provides a fresh look at the evolution of the field, inclusive of the most recent breakthroughs. On the other hand, by analysing the fundamental figures of merit of these circuits, this Element identifies the remaining challenges for the success of organic and amorphous-metal-oxide analogue electronics as the platform of choice for flexible smart-sensor systems.

2 Organic and Amorphous-Metal-Oxide Thin-Film Transistors

2.1 Flexible Thin-Film Transistors

Thin-film transistors (TFTs) are the unit blocks of flexible analogue electronics. In this section we introduce their basic operational principles and geometries, and discuss in detail their parameters and properties relevant to their use in flexible analogue circuits.

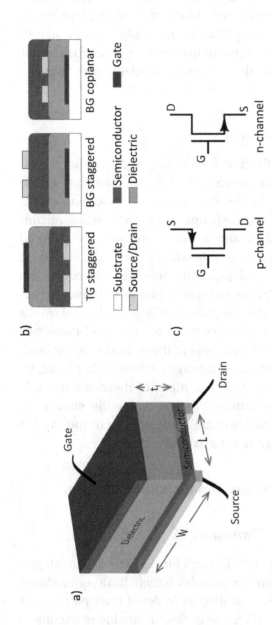

Figure 2.1 a) Basic device structure of a TFT. b) Most common TFT geometries. c) Symbols for p-channel and n-channel TFTs.

2.1.1 Basic Structure and Geometry

A TFT is a three-terminal electronic device whose basic functionality consists in the voltage-controlled modulation of the conductivity of a semiconductor. Figure 2.1a shows its basic structure, which features a semiconducting thin film (typical thickness of less than 100 nm) placed between two electrodes, referred to as *source* and *drain*, running parallel to each other over a distance W, known as the *channel width*, at a distance L from each other, known as the *channel length*. The semiconductor is placed above or below an insulating layer, referred to as the *gate dielectric*, of thickness t_I (typically, $t_I \ll L$). The gate dielectric separates the semiconductor from a third electrode called *gate*, which overlaps the region delimited by source and drain.

Besides the constitutive layers above, TFTs also require a substrate for mechanical support. Depending on the position of the substrate with respect to the semiconductor and electrodes, different geometric implementations of the general TFT structure are possible. Figure 2.1b illustrates the most common geometries. If a TFT has its gate electrode opposite the substrate, it is referred to as *top-gate* (TG); otherwise, it is said to be *bottom-gate* (BG). Additionally, a TFT geometry is called *coplanar* if the source and drain electrodes lie in the same plane of the semiconductor-dielectric interface; otherwise, it is called *staggered*.

2.1.2 Operation and Fundamental Device Parameters

For the purpose of illustrating the basic functionality of a TFT, let us first assume that a positive voltage is applied to the gate ($V_G > 0\,\text{V}$), while the source and the drain are both grounded. Electrons are thus drawn into the semiconductor towards the interface with the gate dielectric (n-channel operation) in a layer only a few nanometres thick (Figure 2.2a) [39]. This effect is called *electron accumulation*, and the electrons induced in the semiconductor are said to constitute the *transistor channel*. If we then establish a potential difference V_{DS} between source and drain, these electrons will start to flow along the semiconductor–dielectric interface, yielding a finite drain-source current I_{DS}. Moreover, by increasing the

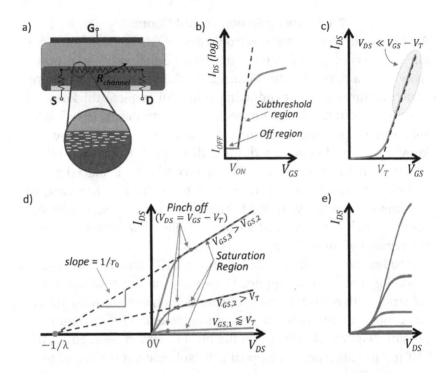

Figure 2.2 a) Schematic depiction of the source-drain resistance, including contact resistance at the source and the drain, and of the conductive channel at the semiconductor–dielectric interface. b)-d) Representative characteristics of an n-channel TFT. b) Transfer characteristics for non-zero V_{DS} (semilogarithmic plot). c) Transfer characteristics in linear scale. d) Output characteristics at different gate voltages, inclusive of the channel-length modulation effect. e) Output characteristics in the presence of significant contact resistance.

magnitude of the gate potential, the channel will become more conductive, and, for a constant V_{DS}, the drain-source current will rise. A symmetric argument holds for negative gate voltages and hole accumulation (p-channel operation).[2] Note that most

---

[2] As can be appreciated from this discussion, TFTs are typically switched on in accumulation, contrary to silicon metal-oxide-semiconductor field-effect transistors (MOSFETs), which conduct large currents in inversion.

semiconductors may only allow either n-channel operation (resulting in n-channel TFTs) or p-channel operation (leading to p-channel TFTs).[3] Depending on whether a TFT is n-channel or p-channel, either of the symbols in Figure 2.1c is used to represent it schematically. The two symbols comprise an arrow that denotes the direction of current flow in accumulation: from source to drain in a p-channel TFT and from drain to source in the n-channel counterpart.

In an actual TFT the gate voltage should be made greater than the so-called threshold voltage V_T for the channel conductance to become appreciable (Figure 2.2c). In fact, to a first approximation the areal charge density in the channel can be expressed as

$$Q_{ch} = C_I (V_{GS} - V_T),$$ (2.1)

where C_I is the areal capacitance of the gate dielectric, and $V_{GS} - V_T$ is the so-called overdrive voltage.

When a TFT is biased above threshold and with a small V_{DS} (specifically, $V_{DS} \ll V_{GS} - V_T$), the conductivity of the channel is approximately uniform. Thus, an increment in V_{DS} yields a proportional increase in drain-source current (linear region of operation), with the proportionality constant being the channel conductance. In this simplified picture, the drain-source current in the linear region reads:

$$I_{DS} \cong \mu C_I \frac{W}{L} (V_{GS} - V_T) V_{DS}$$ (2.2)

where μ is the charge carrier mobility in the channel. If a larger V_{DS} is applied, the channel conductivity will no longer be uniform, as a smaller overdrive voltage will be present between the gate and the points in the channel closer to the drain (Figure 2.2c). Eventually, for $V_{DS} = V_{GS} - V_T$ the overdrive voltage becomes zero at the drain end of the channel. We refer to this specific bias point as *pinch-off*, and to the region of operation for $V_{DS} > V_{GS} - V_T$ as *saturation*. In

---

[3] In a number of cases, both n-channel and p-channel operational modes are possible, leading to ambipolar TFTs.

this region I_{DS} has a very weak dependence on V_{DS}, given the non-uniformity of the channel conductance. In fact, it can be derived that for $V_{DS} > V_{GS} - V_T$ the current becomes approximately independent of V_{DS}:

$$I_{DS,sat} \cong \frac{1}{2}\mu\,C_I\,\frac{W}{L}(V_{GS} - V_T)^2. \tag{2.3}$$

An illustration of the different regions of operation and of the key transistor parameters is given in Figure 2.2 b–d, where I_{DS} is plotted as a function of V_{GS} (transfer characteristics) and as a function of V_{DS} (output characteristics).

While Equation (2.3) suggests that the saturation current is only a function of the gate voltage, in fact it also exhibits a minor dependence on the drain voltage, which is of particular significance for analogue electronics. This effect is due to channel length modulation, i.e., a reduction of the effective channel length as the drain bias increases. A basic model for the saturation current incorporating this effect features an additional term linear in V_{DS}:

$$I_{DS,sat} \cong \frac{1}{2}\mu\,C_I\,\frac{W}{L}(V_{GS} - V_T)^2(1 + \lambda V_{DS}), \tag{2.4}$$

where λ is a parameter that can be graphically understood as the reciprocal of the intercept on the V_{DS} axis of the extrapolation of a transistor's saturated output characteristics (Figure 2.2d). Typically, λ is inversely proportional to the channel length: $\lambda \propto L^{-1}$.

Equations (2.2–2.4) capture to a good approximation the operation of a TFT, yet more refined models become necessary when dealing with semiconductors that exhibit a gate-voltage-dependent charge transport, as is often the case for organic and amorphous-metal-oxide semiconductors. Without delving into the subtleties of the refined models, here it suffices to present one of the most widely used expressions for the saturation current [40], which reads:

$$I_{DS,sat} \cong \frac{1}{2+\gamma} \mu C_I \frac{W}{L} (V_{GS} - V_T)^{2+\gamma} (1 + \lambda V_{DS}).$$ (2.5)

Here μ is still a constant, as the voltage-dependent charge transport is captured by the parameter γ at the exponent of the overdrive voltage.

These equations hold for a transistor whose source and drain contacts allow Ohmic charge injection into the semiconductor. In reality, injection into organic and amorphous-metal-oxide semiconductors may have to overcome a significant energy barrier. This can be modelled as a contact resistance R_c, which lumps contact effects into a resistor in series with both source and drain contacts (Figure 2.2a). To a first approximation, the saturation current would then read [41]:

$$I_{DS,sat} \cong \frac{1}{2+\gamma} \mu C_I \frac{W}{L} (V_{GS} - I_{DS,sat} \cdot R_c - V_T)^{2+\gamma} (1 + \lambda V_{DS}).$$

(2.6)

Contact resistance may have a severe impact on TFT characteristics, typically manifesting itself as a pronounced nonlinearity of the drain current at small V_{DS} (Figure 2.2e). In particular, contact resistance may be especially relevant at low V_{DS}, and may impair altogether low-voltage transistor operation [8]. In the following we assume that contact resistance is negligible, unless stated otherwise.

As a TFT transitions to the *on* (i.e., conducting) state by increasing the gate voltage, an exponential rise in the channel conductivity is observed, and the corresponding region of operation is called the *subthreshold region* (Figure 2.2b). The parameter that quantifies how easily this transition occurs (i.e., how small a change in V_{GS} is necessary for the *off-on* transition to occur) is called *subthreshold slope*, which is defined as:

$$S := \frac{dV_{GS}}{d(\log_{10} I_{DS})}.$$ (2.7)

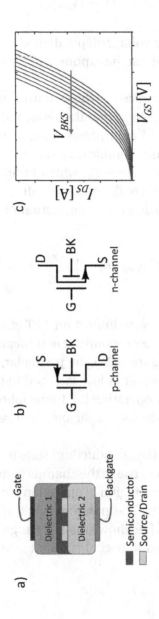

Figure 2.3 Dual-gate TFTs. a) Structure. b) Schematic symbols. c) Effect of backgate voltage on transfer characteristics of an n-channel dual-gate TFT for a backgate voltage below threshold.

The gate voltage V_{ON} from which the exponential rise in channel conductance starts is called the *onset voltage* (Figure 2.2b). For $V_{GS} \leq V_{ON}$, the semiconductor does not deliver a significant current. In this region of operation the TFT is considered *off*, and the corresponding minute drain-source current I_{OFF} is referred to as the *off current*.

2.1.3 Dual-Gate TFTs

A more complex TFT structure that has captured some interest in flexible analogue electronics is obtained by adding an extra gate electrode and gate dielectric layer (see Figure 2.3) to the basic TFT of Sections 2.1.1 and 2.1.2. The resulting device, called *dual-gate* or *double-gate TFT*, is characterised by a similar operation mechanism (i.e., field-based modulation of the channel conductance), but relies on two control terminals to set the channel charge density.

Most often, one gate electrode serves as the primary gate (i.e., it has a much more pronounced impact on the channel current through a stronger capacitive coupling to the semiconductor), while the other (referred to as *backgate*) is used to fine-tune the TFT characteristics. In this case, a dual-gate TFT approximately follows similar equations of a basic TFT (Section 2.1.2). In particular, if the backgate is biased above threshold, the overall channel current is increased by the factor $(1 + \alpha \cdot V_{BKS})$, where α is a coupling constant and V_{BKS} is the voltage between the backgate and the source terminals. If the backgate is biased below threshold, instead, the effect of the backgate can be modelled as a threshold voltage shift $\Delta V_T = -\eta \cdot V_{BKS}$, where η is a modulation coefficient.

2.1.4 Flexibility

While the discussion thus far has referred to generic TFTs, flexible transistors further require one to consider their response under bending, which induces strain in the different layers of the device stack. If subjected to excessive strain, an organic or amorphous-metal-oxide TFT can degrade irreversibly, or even undergo mechanical failure. Therefore, a first important parameter that characterises the flexibility of a TFT is the maximum strain ε_{max} it can endure without undergoing irreversible changes.

Further insight is achieved by considering the dependence of key device parameters on strain. In particular, changes in carrier mobility under moderate strain ε_{TFT} at a bending radius R have been shown to conform to the expression:

$$\frac{\mu(R)}{\mu_0} \cong 1 + m \cdot \varepsilon_{TFT} \tag{2.8}$$

where m is a parameter associated with the given TFT technology, and ε_{TFT} is expressed in percentage points [42]. The parameter m thus reflects the strain sensitivity of a given TFT technology.

A final key point on TFT flexibility concerns the relation between TFT strain and geometrical and mechanical parameters. For a TFT of total thickness d_{TFT} realised onto an inherently flat foil of thickness d_{sub} bent to a radius of curvature R, the strain reads [43]:

$$\varepsilon_{TFT} \cong \frac{1}{R} \frac{d_{TFT} + d_{sub}}{2} \frac{\chi\gamma^2 + 2\chi\gamma + 1}{\chi\gamma^2 + \chi\gamma + \gamma + 1} \cong \frac{1}{R} \frac{d_{sub}}{2}. \tag{2.9}$$

Here, $\chi = Y_{TFT}/Y_{sub}$, Y_{TFT} and Y_{sub} are the Young's modulus of the TFT and the substrate, respectively, and $\gamma = d_{TFT}/d_{sub}$, while the last approximation holds for the typical scenario in which $d_{TFT} \ll d_{sub}$. Equation (2.9) indicates an approximate proportionality between TFT strain and substrate thickness. This implies that TFTs realised on thinner substrates can be bent to a smaller radius (cf. ε_{max}).

2.1.5   Small-Signal Model

Section 2.1.2 describes the behaviour of a TFT subjected to terminal voltages spanning its manifold regions of operation. The corresponding nonlinear equations constitute the so-called large-signal model of a TFT. Analogue circuits, however, most often rely on small perturbations of the terminal quantities around a constant set (I_{DS}, V_{DS}, V_{GS}) (known as *direct-current (DC) bias point*). It is therefore useful to introduce the so-called small-signal

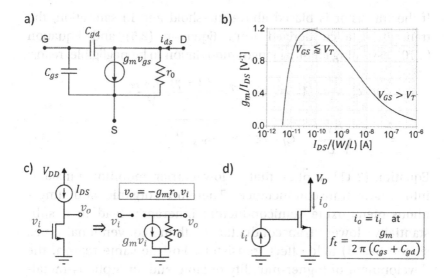

Figure 2.4 a) Small-signal model of a TFT. b) g_m/I_{DS} versus $I_{DS}/(W/L)$ for a generic organic TFT. c) Derivation of intrinsic gain. d) Derivation of transition frequency.

model of a TFT, which is based on the linearisation of the device behaviour around its bias point.

A typical simple small-signal model of a TFT is shown in Figure 2.4a. The small-signal quantities are here expressed with symbols in lower case (both variable names and subscripts), while the bias point is shown with all capital letters. This model represents a TFT as a voltage-controlled current source providing a current equal to the product $g_m \cdot v_{gs}$. The transconductance g_m of a TFT is defined as[4]:

$$g_m := \frac{\partial I_{DS}}{\partial V_{GS}}\bigg|_{at\,bias\,point}. \tag{2.10}$$

---

[4] The definition of transconductance, as the one of output resistance (Equation (2.16)), is generally given by introducing the total signal, sum of the small-signal and the bias-point components. For the sake of simplicity, however, here we make no distinction between total signal and bias-point quantities.

If the transistor is biased above threshold and in saturation, the transconductance derived from Equation (2.5) and Equation (2.10), assuming channel length modulation to be negligible, reads:

$$g_m = \mu \cdot C_I \cdot \frac{W}{L} \cdot (V_{GS} - V_T)^{1+\gamma} = \frac{(2 + \gamma) \cdot I_{DS,sat}}{V_{GS} - V_T}$$

$$= \sqrt[(2 \cdot +\gamma)]{\frac{W}{L} \cdot \mu \cdot C_I \cdot [(2 + \gamma) \cdot I_{DS,sat}]^{1+\gamma}}. \tag{2.11}$$

Equation (2.11) implies that a low carrier mobility translates into a low transconductance. Therefore, organic and amorphous-metal-oxide semiconductors inherently lead to a substantially lower transconductance than conventional (i.e., silicon-based) field-effect transistors. For the same reason, the development of higher-mobility organic and amorphous-metal-oxide semiconductors is expected to enable higher transconductance figures.

In the presence of contact effects, as often found in organic and amorphous-metal-oxide TFTs, the transconductance can no longer be expressed as in Equation (2.11), because the saturation current does not obey Equation (2.5). Here it suffices to mention that, in this case, the transconductance can be expressed as:

$$g_m = \frac{(2 + \gamma) \cdot I_{DS,sat}}{V_{GS} - V_T + (1 + \gamma) \cdot I_{DS,sat} \cdot R_c} \tag{2.12}$$

(cf. [41]). Algebraic manipulation and comparison with Equation (2.11) indeed reveals that contact resistance leads to a reduction of the TFT transconductance:

$$g_m = \frac{g_{m0}}{1 + g_{m0} \cdot R_c}. \tag{2.13}$$

Here g_{m0} is the transconductance for zero contact resistance, i.e., calculated as in Equation (2.11). The derivation of Equation (2.12) and Equation (2.13) is detailed in the Supplementary Information available at www.cambridge.org/pecunia.

The relationship between the ratio g_m/I_{DS} and the normalised drain-source current $\tilde{I}_{DS} = I_{DS}/(W/L))$ has an important role in analogue circuit design [44]. In particular, g_m/I_{DS} is a measure of the efficiency of a TFT to translate its current (hence dissipated power) into transconductance. The greater g_m/I_{DS} is, the higher the transconductance obtained at a given bias current. Analytically, this quantity can be expressed as:

$$\frac{g_m}{I_{DS}} = \frac{1}{I_{DS}} \cdot \frac{\partial I_{DS}}{\partial V_{GS}} = \frac{\partial(\ln I_{DS})}{\partial V_{GS}} = \frac{\partial}{\partial V_{GS}} \left[\ln \left(\frac{I_{DS}}{W/L} \right) \right]. \tag{2.14}$$

Inspection of Equation (2.14) makes it apparent that g_m/I_{DS} is maximum in subthreshold ($V_{GS} \lesssim V_T$), where the dependence $I_{DS} - V_{GS}$ is exponential (cf. Figure 2.2b), whereas it decreases as the bias point moves above threshold ($V_{GS} > V_T$). For the sake of illustration, the experimental ratio g_m/I_{DS} of an organic TFT is presented in Figure 2.4b, making clear the trend discussed earlier. Additionally, Equation (2.14) indicates that the ratio g_m/I_{DS} is not significantly impacted by the mobility level of a given semiconductor technology. In fact, in saturated accumulation region[5]:

$$\frac{g_m}{I_{DS}} \cong \frac{\partial}{\partial V_{GS}} \left[\ln(\mu) + \ln \left(\frac{C_I}{2+\gamma} (V_{GS} - V_T)^{2+\gamma} \right) \right]$$

$$\cong \frac{\partial}{\partial V_{GS}} \left[\ln \left(\frac{C_I}{2+\gamma} (V_{GS} - V_T)^{2+\gamma} \right) \right]. \tag{2.15}$$

This explains why, for instance, some organic TFT technologies can approach the g_m/I_{DS} ratios of conventional silicon electronics above threshold [45].

As shown in Figure 2.4a, the small-signal model also includes a resistor r_0, referred to as the *TFT output resistance*, which captures the effect of channel length modulation (cf. Figure 2.2d). Its definition reads:

---

[5] With *saturated accumulation* we indicate the biasing region where a TFT is in accumulation and saturation.

$$\frac{1}{r_0} := \frac{\partial I_{DS}}{\partial V_{DS}}\Big|_{at \, bias \, point}.$$ (2.16)

Plugging in Equation (2.5), we obtain:

$$\frac{1}{r_0} = \lambda \cdot \frac{W}{L} \cdot \frac{\mu \cdot C_I}{(2+\gamma)} \cdot (V_{GS} - V_T)^{2+\gamma} = \lambda \cdot I_{DS,sat} \cong \lambda \cdot I_{DS}.$$ (2.17)

The final approximation is based on the assumption that λ is so small that $I_{DS} \cong I_{DS,sat}$.

The small-signal model also accounts for the frequency-dependent behaviour of a TFT via the capacitors C_{gs} and C_{gd}, the so-called gate-source and gate-drain capacitances. C_{gs} is primarily determined by the channel capacitance, which is approximately equal to $2/3 \cdot W \cdot L \cdot C_I$ in the saturation region. An additional term arises from the physical overlap between source and gate contacts (Figure 2.1a) and the fringing capacitance. The total gate-source capacitance in saturation is thus given by

$$C_{gs} = \frac{2}{3} \cdot W \cdot L \cdot C_I + W \cdot L_{ov,s} \cdot C_I,$$ (2.18)

where $L_{ov,s}$ is the effective overlap distance for the source, usually derived empirically. Conversely, the channel capacitance does not contribute to C_{gd} in the saturation region. Therefore, C_{gd} is primarily determined by the overlap between the gate and the drain and the fringing capacitance:

$$C_{gd} = W \cdot L_{ov,d} \cdot C_I,$$ (2.19)

where $L_{ov,d}$ is the effective overlap distance for the drain. Typically, in saturation $C_{gs} \gg C_{gd}$.

2.1.6 Electronic Noise

The discussion thus far has dealt with the deterministic behaviour of TFTs. A complete description of TFT operation, however, requires one to consider also the random fluctuations (i.e., noise) in terminal currents and voltages. This effect is most often described as an extra (random) component in the drain-source current. At sufficiently low frequencies, its power spectral density typically presents an inverse dependence on frequency (flicker noise). This is in fact the component that dominates in the operation region of interest for flexible TFTs, in view of their limited bandwidth (see Section 2.1.7). The power spectral density of the drain-source flicker-noise current often conforms to Hooge's phenomenological expression

$$S_{IDS}(f) = \alpha_H \frac{I_{DS}^2}{Nf}, \tag{2.20}$$

where f is the frequency, α_H is the Hooge parameter and N is the total number of free charge carriers. The Hooge parameter is thus an important figure of merit to evaluate the noise behaviour of different TFT technologies.

According to classical analysis, two possible mechanisms contribute to flicker noise: fluctuation of the number of carriers under the gate and bulk mobility fluctuation. The increased channel mobility that characterises modern TFTs is making the injection processes between the source and the channel more and more relevant. The impact of these injection mechanisms on the noise behaviour has not been investigated in detail yet.

2.1.7 Key Figures of Merit for Analogue Circuit Applications

While large-signal characteristics and small-signal parameters provide a complete representation of transistor behaviour, it is advantageous for analogue circuit design to define a reduced list of parameters capable of highlighting key aspects of transistor performance for a given technology. These parameters can also be used as figures of merit when comparing the transistor performance of different technologies. In particular, key reduced analogue parameters of a TFT are its intrinsic gain and transition frequency, as discussed next.

Intrinsic Gain

Transistors are the building blocks of analogue amplifiers, i.e., circuits that boost the signals at their input by a factor called *amplification gain*. While amplification is discussed at length in Chapter 3, here it suffices to note that, conceptually, the simplest and smallest amplifier can be realised with just one transistor, if configured as in Figure 2.4c. The ratio between the small-signal output and input voltages of this circuit defines the transistor's intrinsic gain A_{Vi}. Directly from the small-signal model (see Figure 2.4c), it can be found that

$$A_{Vi} = \left| \frac{v_o}{v_i} \right| = g_m \cdot r_0. \tag{2.21}$$

Therefore, a transistor's intrinsic gain is set by its output resistance and transconductance. This is the highest gain achievable by an amplifier in this configuration, hence it constitutes a key gain value for a given transistor technology.

Plugging in the expressions of the transconductance and the output resistance (i.e., Equation (2.11) and Equation (2.17)), we obtain that the intrinsic gain in saturated accumulation is approximately given by:

$$A_{Vi} \cong \frac{g_m}{I_{DS,sat}} \cdot \frac{1}{\lambda} = \frac{2 + \gamma}{\lambda \cdot (V_{GS} - V_T)}. \tag{2.22}$$

Equation (2.22) provides two important guidelines for the design of TFTs with large intrinsic gain: a) they should have a long channel, since $\lambda \propto L^{-1}$; b) when biased in accumulation, they should operate with small overdrive voltage.

Transition Frequency

The transition frequency f_t of a transistor is indicative of its intrinsic speed. It is a key figure of merit because it has a direct impact on the speed of resulting analogue circuits. It is defined as the frequency at which a transistor, connected as in Figure 2.4d, presents a small-signal current gain equal to unity:

$$A_I = \left| \frac{i_o}{i_i} \right| = 1 \cong \frac{g_m}{2 \cdot \pi \cdot f_t \cdot \left(C_{gs} + C_{gd} \right)}. \tag{2.23}$$

Plugging in the expression of g_m in saturated accumulation (Equation (2.11)), we obtain:

$$f_t \cong \frac{g_m}{2 \cdot \pi \cdot \left(C_{gs} + C_{gd} \right)} = \frac{\mu \cdot C_I \cdot W \cdot \left(V_{GS} - V_T \right)^{1+\gamma}}{2 \cdot \pi \cdot L \cdot \left(C_{gs} + C_{gd} \right)}. \tag{2.24}$$

Equation (2.24) indicates that large physical overlap between source/drain and gate contacts would severely limit transistor speed. Assuming that overlap capacitances are negligible and using the common approximations for the channel capacitance in accumulation (Equation (2.18)), it follows that:

$$f_t \cong \frac{3 \cdot \mu \cdot \left(V_{GS} - V_T \right)^{1+\gamma}}{4 \cdot \pi \cdot L^2}. \tag{2.25}$$

If high-speed operation is required, the design strategy must aim to use transistors in accumulation with minimum channel length and large overdrive voltages. This is clearly in contrast with the design criteria for maximum intrinsic gain. Consequently, a general trade-off exists between gain and speed in analogue circuits.

2.2 Semiconductors for Flexible Analogue Electronics

In this section we present the two classes of semiconductors that have recently attracted significant attention for their use in flexible analogue electronics: *organic semiconductors* (OSs) and *amorphous-metal-oxide semiconductors* (AMOxSs). Specifically, here we introduce their basic properties that are relevant to the rest of this work. The interested reader is referred to [46] for OSs and to [47], [48] for AMOxSs to find a more comprehensive treatment of these materials.

2.2.1 Organic Semiconductors

Organic semiconductors (OSs) constitute a vast class of carbon-based materials, comprising both small molecules and polymers.

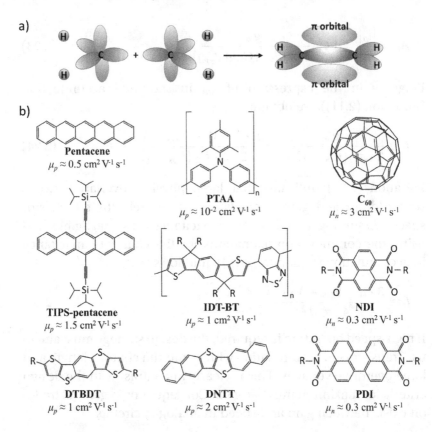

Figure 2.5 a) Molecular orbital depiction of an ethene molecule. b) Chemical formulas and approximate mobility of notable OSs for flexible analogue electronics. N-channel mobility is indicated as μ_n, while μ_p denotes p-channel mobility.

Typical mobility values for thin films of the most recently synthesised OSs are in the range of $1 - 10 \, cm^2 \, V^{-1} \, s^{-1}$. Their common feature is a π-conjugated structure, namely the alternation of single and double covalent bonds. It is the delocalised nature of the associated π and π* molecular orbitals (shown in Figure 2.5 for the simple case of an ethene molecule) that results in the semiconducting properties of these materials.

Conjugation length (i.e., the spatial extension of the delocalised orbitals) in polymer semiconductors is limited by kinks and defects along the polymer chain, hence their π and π^* orbitals result in two quasi-continuous energy bands separated by an energy gap ($\approx 1 - 3$ eV). The highest occupied molecular orbital is generally indicated by its acronym, HOMO. Similarly, the lowest unoccupied molecular orbital is referred to as LUMO. By extension, the bands of π and π^* orbitals are referred to as HOMO and LUMO bands. Equivalent concepts apply to small-molecule OSs.

The conductivity of an organic semiconductor can be modulated by filling the LUMO band or by empting the HOMO one by some doping mechanism. Indeed, OSs are capable of both n-channel and p-channel conduction, and sometimes the two coexist in the same material.[6] However, while p-channel behaviour is widespread and often in the high-mobility range, only in the past few years have n-channel materials started catching up in performance. This performance gap has been generally attributed to the fact that the charge trapping of electrons in the LUMO band is more severe than trapping processes which holes in the HOMO are subject to [49].

For many years, the performance benchmark for solution-processed p-channel polymers was set by P3HT and PBTTT [50], [51]. The transport properties of these polymers rely on an engineered crystalline microstructure, which leads to a mobility up to $\approx 0.3\,\text{cm}^2\,\text{V}^{-1}\,\text{s}^{-1}$. Other very popular polymers were PTAAs (e.g., see Figure 2.5b), whose attractiveness is linked to their amorphous character and associated low sensitivity to process variations. More recently, donor-acceptor copolymers have come to the fore (e.g., indacenodithiophene-based copolymers such as IDT-BT, see Figure 2.5b), delivering mobilities above $1\,\text{cm}^2\,\text{V}^{-1}\,\text{s}^{-1}$ [52], [53]. In contrast to earlier top-performance systems (e.g., PBTTT), these mobility values are achieved with a much less ordered microstructure (in some cases, near amorphous), thus are less sensitive to process variations.

---

[6] Ambipolar OSs have limited applicability in analogue electronics, thus no further mention is made of them in the remainder of this work.

As for small-molecule semiconductors, pentacene and its solution-processible derivatives (e.g., TIPS-pentacene) constituted a long-standing benchmark for p-channel transport, giving field-effect mobilities up to $1 - 2\,cm^2\,V^{-1}\,s^{-1}$ (Figure 2.5b). More recently, thienoacenes, i.e., acenes incorporating thiophenes in their structures (e.g., BTBT, DNTT, DTBDT, see Figure 2.5b), have led to higher carrier mobility, in excess of $5\,cm^2\,V^{-1}\,s^{-1}$ in some cases.

While n-channel small-molecule OSs had long suffered the comparison with their p-channel counterpart, a major breakthrough was achieved with diimide derivatives, most notably naphthalene- and perylene-based diimides (NDIs and PDIs, respectively, as shown in Figure 2.5b). N-channel mobilities up to approximately $1\,cm^2\,V^{-1}\,s^{-1}$ have been observed in solution-processed NDI and PDI transistors [54]. As for polymers, NDI-based P(NDI2OD-T2) is a long-standing high-performance benchmark, with typical mobility in the $0.1 - 0.6\,cm^2\,V^{-1}\,s^{-1}$ range [55]. More recently, n-channel diketopyrrolopyrrole-based polymers have surpassed these performance levels, leading to values greater than $1\,cm^2\,V^{-1}\,s^{-1}$ in some cases [54]. In spite of these outstanding developments, the stability of n-channel OSs under atmospheric operating conditions

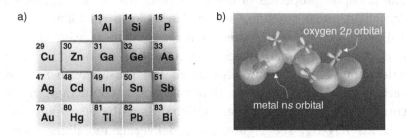

Figure 2.6 Metal-oxide semiconductors. a) Typical metal components and their location in the periodic table (elements within the contour). b) Spatial arrangement (in an amorphous structure) of the elemental orbitals contributing to the conduction and valence bands – adapted with permission from Macmillan Publishers Ltd: Nature ([59]). Copyright (2004).

Table 2.1 Comparison of key performance parameters of organic and metal-oxide semiconductors with benchmark technologies.

	a-Si	poly-Si (Low-T)	AMOxSs	OSs
Mobility (cm^2 V^{-1} s^{-1}) [46][48][60][68][69]	< 1	50 – 100	1 – 50	1 – 10
Subthreshold Slope (V dec^{-1}) [60]	0.4 – 0.5	0.2 – 0.4	0.1 – 0.6	0.1 – 1
Threshold Voltage (V) [48][68]	1 – 2	1 – 2	0 – 2	0 – 20
Noise: α_H [62]-[67]	$3 \cdot 10^{-3} - 5 \cdot 10^{-1}$	$6 \cdot 10^{-1}$	$4 \cdot 10^{-4} - 1$	$2 \cdot 10^{-2} - 1$
Channel Polarity	n	n and p	n	n and p
Reliability	moderate	high	high	moderate/high
Process Temperature (°C)	250 – 350	< 500	RT – 250	RT – 100
Flexibility [42]: ε_{max} (%)	0.2 – 0.4	0.1 – 0.5	0.7	1 – 2
$|m|$	20 – 30	20 – 40	1.4 – 2	5 – 15

still represents an issue, somehow hindering the adoption of n-channel OSs in flexible electronics.

2.2.2 Amorphous-Metal-Oxide Semiconductors

Amorphous-metal-oxide semiconductors (AMOxSs)[7] constitute a class of amorphous materials possessing excellent semiconducting behaviour while allowing low processing temperatures (down to room temperature (RT)). They comprise single and multicomponent oxides of a variety of metals with an $(n-1)d^{10}ns^2$ electronic configuration ($n \geq 5$), the most notable being oxides of In, Zn, Ga, Sn (Figure 2.6a). These materials exhibit high electron mobility $(1-50\,\mathrm{cm^2\,V^{-1}\,s^{-1}})$ and operational stability, making them suitable for a wide range of applications. Additionally, their large bandgap $(E_g \geq 3\,\mathrm{eV})$ makes them transparent in the visible, thus enabling transparent electronics.

Metal-oxide semiconductors are characterised by chemical bonding with strong ionic character. In particular, their valence band originates from the oxygen $2p$ orbitals, and their conduction band arises from the metal ns orbitals [56]. A depiction of these orbitals is given in Figure 2.6b. Given its s nature, the conduction band of an AMOxS is largely insensitive to spatial disorder. Indeed, the s orbitals of the metal cations in an AMOxS not only possess spherical symmetry but also have very large radiuses thanks to their high principal quantum number ($n \geq 5$), as shown in Figure 2.6b. Therefore, the spatial overlap of these orbitals is not disrupted by the disorder of bond angles and lengths, and gives rise to a conduction band with a large dispersion and a sharp tail. Conversely, the valence band originates from the oxygen p orbitals, which are more compact and spatially directional. The overlap of these orbitals is thus extremely sensitive to the distortions of the metal–oxygen bonds, yielding a valence band composed of strongly localised states centred on the oxygens [57]. In terms of charge transport, the large overlap of the metal s orbitals

---

[7] For the sake of simplicity, in the following we always take for granted their amorphous nature, and refer to them as *metal-oxide semiconductors*.

and the small density of conduction tail-state density results in a large electron mobility, whereas the localised nature of the valence band states inhibits hole transport.

In recent years, substantial efforts have been made to develop metal-oxide semiconductors capable of p-channel behaviour, e.g., copper oxides, copper-bearing oxides and tin oxide [58]. In spite of some encouraging results, these materials typically yield mobility values $\lesssim 1\,cm^2\,V^{-1}\,s^{-1}$ and are characterised by a generally high defect state concentration [58]. Therefore, p-channel metal-oxide semiconductors are still lagging behind in performance and stability with respect to their n-channel counterpart, and new material design and processing strategies are needed to address this challenge.

2.2.3 Benchmark Semiconductor Technologies

Table 2.1 presents a comparison of organic and metal-oxide semiconductors with other technologies relevant to flexible electronics. The latter are based on amorphous silicon (a-Si) and polycrystalline silicon (poly-Si) [60]. a-Si can be deposited at relatively low temperatures ($\approx 250°\,C$) and over large areas (by plasma-enhanced chemical vapour deposition (PECVD)), and it has been successfully employed in displays and photovoltaics. a-Si suffers from low mobility ($< 1\,cm^2\,V^{-1}\,s^{-1}$) and bias-stress instability, however, which makes it unsuitable for applications demanding continuous operation at large currents [46]. On the other hand, poly-Si has a substantially larger charge-carrier mobility ($50 - 100\,cm^2\,V^{-1}\,s^{-1}$), but typically requires much higher processing temperatures ($400 - 500\,°C$) and a complex fabrication method (involving laser annealing of a-Si precursor layers) burdened with uniformity issues.

Metal-oxide semiconductors are superior in all respects to benchmark technologies: they do not suffer from the low mobility and instability of a-Si, and they can be produced at much lower temperatures than poly-Si. However, these excellent properties manifest only in n-channel operation, given that efforts towards high-performance p-channel oxide semiconductors are still under way. As for OSs, they allow p-channel performance unequalled by AMOxSs, and well above that of a-Si. Their n-channel performance is generally inferior to AMOxSs, however. In terms of stability, OSs have been traditionally inferior to AMOxSs, yet recent studies have demonstrated that they can also attain excellent stability [61].

Flicker noise of all technologies covers a very broad range [62]–[67]. The limited number of studies on the topic makes it impossible to conclude on how these technologies compare with one another in terms of noise behaviour. It is apparent, however, that TFT flicker noise is worse by at least a factor of ten compared to conventional IC technology [64].

Finally, organic and metal-oxide TFTs have significantly greater flexibility compared to benchmark technologies [42]: their maximum bearable strain ε_{max} is nearly an order of magnitude higher than that of a-Si or poly-Si, OSs being the best according to this metric; their relative change in carrier mobility under strain is well below that of a-Si or poly-Si (cf. $|m|$ values in Table 2.1), with AMOxSs being the most insensitive. This clearly evidences that AMOxSs and OSs are extremely attractive TFT technologies for flexible electronics.

2.2.4　Key Analogue Figures of Merit of Organic and Metal-Oxide TFTs

Having discussed their general properties, we now examine the suitability of organic and metal-oxide semiconductors for analogue electronics with respect to the key figures of merit introduced in Section 2.1.7.

While the transconductance of organic and metal-oxide TFTs is bound to be substantially lower than conventional (silicon-based) analogue electronics due to their comparatively poor charge transport, the intrinsic gain achievable with these flexible technologies is, in fact, independent of their low carrier mobility (cf. Equation (2.21)). This is an important point, in particular considering the numerous controversial statements on this aspect in the literature. While many authors refer to OSs as being affected by poor intrinsic gain, in actuality, the few works that present experimental evaluation of this key figure of merit have reported values up to 140 V/V for pentacene-based TFTs [70], [71], and in the 200 – 400 V/V range for DNTT and $F_{16}CuPc$ TFTs [45]. As for metal-oxide TFTs, values as high as 165 V/V have been measured [72]. To put things in perspective, conventional analogue electronics, as realised with modern short-channel silicon technology, is characterised by intrinsic gains of 40 V/V or less [73]. This obviously indicates that organic and metal-oxide semiconductors are surely suitable for analogue applications. As a cautionary note, contact resistance is a recurrent issue in flexible TFTs, especially organic ones, and, as pointed out in Section 2.1.5, contact resistance has a negative impact on transconductance. It is therefore imperative to minimise contact resistance for a given organic/metal-oxide technology to exploit its full analogue potential.

Charge carrier mobility is an essential element determining the TFT transition frequency (cf. Section 2.1.7). Therefore, organic and metal-oxide TFT technologies obviously cannot approach the bandwidth figures of conventional (silicon-based) analogue electronics. Nonetheless, experimentally evaluated transition frequencies evidence the compatibility of these flexible TFT technologies with a wealth of analogue applications: f_t values at around 10 – 70 kHz were obtained from organic transistors optimised for analogue circuits [45], [70], [71], while metal-oxide semiconductors can reach f_t values well beyond tens of megahertz [72].[8] Such frequency ranges are certainly well suited for smart-sensor system applications, which typically involve signals within the 10 – 20 kHz range.

---

[8] It is noteworthy that much higher transition frequencies have been reported in the general literature on organic and metal-oxide TFTs (tens of M and > 100 M, respectively) [153], [154]. Such impressive values, however, are obtained with considerably short channel lengths ($\lesssim 1$ μm), which generally lead to degraded intrinsic gain values (approaching 0 dB [72]). This gain–speed trade-off, already captured in the discussion of Equation (2.22) and Equation (2.25), makes it apparent that analogue circuits do not generally exploit the maximum speed of a technology. Hence, the conservative transition frequencies provided here, referred to organic and metal-oxide transistors for analogue applications, are the most representative values in the context of flexible analogue electronics.

2.3   Gate Dielectrics for Flexible Analogue Electronics

The gate dielectric is a key component of a TFT: not only does it provide electrical insulation between semiconductor and gate electrode but it also contributes to the properties of the interface along which charge transport occurs.

To be an effective gate dielectric, an insulator is required to confine the channel charge to the semiconductor it forms an interface with. This sets a lower limit to the band offsets between the semiconductor and the gate dielectric, which should be greater than approximately 1 eV [74]. Additionally, a good gate dielectric must be free of interfacial trap states, which are detrimental to device performance and stability.

A large number of insulators have found use as gate dielectrics in organic and metal-oxide TFTs. According to their composition, they can be classified as inorganic, organic and nanostructured. Typical inorganic dielectrics are binary oxides such as SiO_2, Al_2O_3, ZrO_2 and HfO_2. As for organic dielectrics, they are most often polymers lacking conjugation and residual ionic species. Representative examples are polystyrene (PS), poly(methyl methacrylate) (PMMA) and poly(4-vinyl phenol) (P4VP, see Figure 2.7c). Nanostructured dielectrics, instead, come in the form of a matrix (typically polymeric) that incorporates inorganic nanoparticles (such as TiO_2 [75], [76] and $BaTiO_3$ [77], [78]).

The charge accumulated in a TFT channel is proportional to the areal capacitance C_I of its gate dielectric (Equation (2.1)). For a linear dielectric with uniform thickness t_I, C_I can be expressed as $\epsilon_0 \kappa / t_I$, where ϵ_0 is the vacuum permittivity and κ is the relative permittivity of the dielectric material. With respect to their permittivity values, it is customary to refer to gate dielectrics as low-κ or high-κ depending on whether their relative permittivities are smaller or larger than that of ubiquitous silicon dioxide ($\kappa = 3.9$).

Figure 2.7b shows the relative permittivity of common organic and inorganic gate dielectrics. Most organic dielectrics are low-κ; quite a few manifest high-κ behaviour, as allowed by the presence of polar groups (e.g., hydroxyl, nitrile and carbonyl), but tend to be

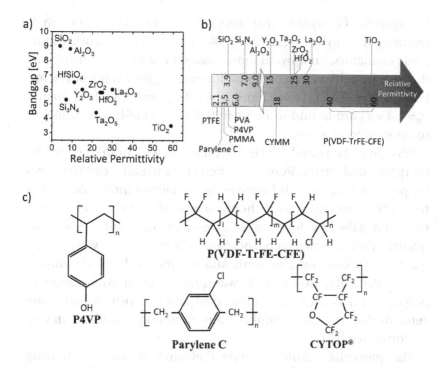

Figure 2.7 a) Bandgap versus relative permittivity of common inorganic dielectrics, evidencing an inverse trend. b) Comparison of relative permittivity values of common organic and inorganic dielectrics (bulk values). c) Chemical formulas of notable polymer dielectrics employed in flexible analogue electronics.

hygroscopic and result in TFTs susceptible to instability. An exceptional case is that of solution-processable PVDF-based terpolymers (e.g., P(VDF-TrFE-CFE), see Figure 2.7c), which are relaxor ferroelectric materials that attain a κ as high as $40-60$ while maintaining a quasilinear behaviour up to fields of several $MV\,cm^{-1}$ [79]. Inorganic dielectrics, instead, populate the relative permittivity axis more uniformly.

Few data are available on the frontier energy bands of organic insulators, while those of inorganic insulators have been extensively characterised. Specifically, it has been found that the bandgap E_g of inorganic insulators is inversely dependent on their relative permittivity (Figure 2.7a) [80]. Therefore, when using inorganic insulators, an upward limit exists on the relative permittivity that can be achieved in combination with a given semiconductor. For instance, Ta_2O_5 ($\kappa = 22$ and $E_g = 3.2$ eV) and TiO_2 ($\kappa = 80$ and $E_g = 4.4$ eV) are bound to be unsuitable for wide-bandgap metal-oxide semiconductors.

OSs have been employed in combination with both organic and inorganic dielectrics. Polymeric dielectrics usually lead to superior performance. This is because the surface of inorganic dielectrics often terminates with chemical groups that act as electronic traps for OSs. As for metal-oxide-semiconductor TFTs, they feature almost exclusively inorganic dielectrics, in view of the purportedly superior electronic and thermomechanical compatibility. Recently, however, it was found that a wide range of polymer dielectrics are also compatible with metal-oxide semiconductors, this combination attaining excellent device performance [81].

The particular choice of gate dielectric is key to achieving low-voltage transistor operation (maximum terminal voltage below ≈ 3 V), which is especially challenging for solution-processed semiconductors (see Section 2.5.1). Low-voltage operation in organic and metal-oxide TFTs has been usually achieved via thinner [82], [83] and higher-κ [81], [84]–[86] dielectrics. A notable extreme case is that of gate dielectrics consisting of self-assembled monolayers with thickness of about 2 nm, allowing transistor operation below 1 V [87], [88]. Thin and high-κ dielectrics enable low-voltage operation of metal-oxide-semiconductor TFTs with relative ease, thanks to the excellent electronic properties of AMOxSs [81]. In the case of OSs, however, the desired low-voltage operation does not scale linearly with relative permittivity, as more polar dielectrics degrade organic semiconductor mobility [89], [90]. Moreover, the presence of trapping effects may pose a severe obstacle to their low-voltage operation, and trap healing techniques may become necessary [8].

2.4 Substrate Materials for Flexible Analogue Electronics

Ideal substrates for flexible electronics are tough polymeric films, able to withstand both the temperatures and the chemicals required for all circuit fabrication steps. Polymeric materials that have been successfully demonstrated as substrates for flexible electronics include polyethylene terephthalate (PET), polyethylene naphthalate (PEN) and polyimide (PI). All three have excellent mechanical properties and chemical resistance. PI allows the greatest thermal budget (up to ≈ 300 °C, in contrast to ≈ 180 °C for PEN and ≈ 130 °C for PET), at the price, however, of a greater coefficient of thermal expansion. For sufficient flexibility (i.e., bending radius below 1 cm), foils made of any of these materials should have a thickness smaller than approximately 200 μm.

Substrate flexibility not only enables new form factors to the benefit of end applications but may also allow greater convenience at the manufacturing level. Indeed, producing electronics on cheap plastic substrates is potentially more cost-effective than on rigid silicon or glass, in terms of both material cost and maximum achievable throughput. In particular, flexible substrates lend themselves to roll-to-roll (R2R) circuit fabrication, which involves processing on rolls of foil up to several kilometres long and a few metres wide. In such a production line, circuit fabrication on foil would be performed while the substrate is unwound and seamlessly transferred through the different layer-deposition and layer-patterning stages, potentially enabling high throughput.

Recent years have witnessed a continuous push for flexible electronics on ever thinner substrates. Impressive demonstrations of electronics on foils only a few micrometres thick have been produced [8], [27], [91], [92], in many instances being capable of circuit functionality down to submillimetre bending radiuses. In view of the associated outstanding lightness ($\approx$ g m^{-2}), this approach has been termed *imperceptible electronics*.

2.5   Processing Methods for Flexible Analogue Electronics

A wide range of deposition methods has been explored for flexible analogue electronics. The most fundamental classification distinguishes between vacuum-based and solution-based methods, along with their capability to inherently deposit patterned structures, i.e., additive versus subtractive methods. This section elaborates on these concepts and summarises some of the developments in this area.

2.5.1   Vacuum-Based versus Solution-Based Methods

The first to be employed in organic and metal-oxide electronics, vacuum-based methods are either inherited from or closely related to conventional silicon technology. Among them we find vacuum sublimation of organic (small-molecule) semiconductors, and RF-magnetron sputter coating of metal-oxide semiconductors. Inorganic gate dielectrics are also frequently deposited by vacuum deposition methods, such as PECVD [93], [94], sputter coating [95], [96] and atomic layer deposition (ALD) [97]. These techniques generally afford greater control of composition and microstructure than solution-based methods. They may present important limitations, however: for instance, PECVD requires a high process temperature, conventional ALD is characterised by low throughput, and sputter coating involves energetic ions that may damage the underlying device stack. All these techniques have been or are in the process of being adapted to R2R production, which may come, however, with high-cost investment and maintenance. Indeed, intrinsically vacuum-based techniques such as PECVD and sputter coating involve either complex substrate handling (roll rewinding, unloading, reloading and roll pumping) or a complex arrangement to allow a seamless air-to-vacuum transfer (a series of chambers at intermediate pressures). As for ALD, its adaptation to atmospheric-pressure and low-temperature R2R production is conceptually possible and currently under investigation, but so far the films produced are substantially less dense than their high-temperature vacuum-processed counterparts [98], [99].

Solution-based processing marks a sharp divide between silicon electronics and organic and metal-oxide technologies. Solution-based methods involve the dissolution of the electronic materials or their precursors in suitable solvents. The resulting inks can then be deposited in manifold ways, e.g., via printing and coating. These methods can be carried out at atmospheric pressure and at room temperature. If the ink comprises the electronic materials in their final form, an annealing step at moderate temperature ($T_p < 100\,°C$) is usually required after deposition simply to drive off the residual solvents. If the ink is precursor-based, instead, a curing step at a higher temperature ($T_p \approx 150 - 300\,°C$) is needed to activate the reaction leading to the desired electronic materials. These methods are applicable to OSs, AMOxSs, and organic/inorganic dielectrics. One important requirement relevant to solution-based deposition is that the ink composition does not affect the materials already present on the substrate, i.e. does not dissolve or swell any of the underlying layers. All this considered, solution-based methods allow an extraordinary simplification of the manufacturing process and inherently lend themselves to R2R production, potentially leading to a substantial reduction in capital expenditure and maintenance costs.

While extremely attractive from a manufacturing point of view, solution-based processing is not free of challenges. Indeed, solution-processed semiconductors generally present a higher density of trap states, due to the greater structural disorder and/or higher impurity levels inherent in solution processing. This implies that solution-processed organic and metal-oxide semiconductors generally present lower mobility values, shallower sub-threshold slopes, and larger threshold voltages, making it more challenging, for instance, to achieve low-voltage TFT operation. All this does not jeopardise the high technological appeal of solution processing, however. Therefore, the challenges associated with solution processing have motivated intense research efforts, which, in recent years, have enabled solution-processed semiconductors with performance levels approaching the vacuum-deposited counterparts.

2.5.2   Additive versus Subtractive Processing

Subtractive deposition methods involve blanket deposition on the substrate, and therefore require the selective removal of some of the deposited material to achieve the desired final coverage. This is typically carried out within a multistep photolithography-based process flow. Additive methods, instead, involve the deposition of a material only at desired substrate locations, thus directly leading to the creation of patterned structures. It is thus apparent that additive methods are capable of streamlining the manufacturing process, potentially leading to higher throughput and cost reduction.

Vacuum-based deposition methods typically require subtractive processing, while the solution-based counterparts can be carried out in additive mode. One of the most explored examples of additive processing is inkjet printing, which is a non-contact method involving the on-demand deposition of an electronic-material ink on substrate regions preassigned by design. While conventional inkjet printing methods have a relatively coarse resolution ($\approx$ 20 − 50 μm), recent advances in inkjet printing technology have brought resolution figures down to $\approx$ 1 μm [100], thus comparable to photolithography. Other approaches that have attracted significant attention for TFT fabrication are contact printing methods (e.g., offset, flexographic and gravure printing), where the desired pattern is transferred onto the substrate through physical contact with an inked patterned structure [101]. These approaches are particularly attractive in view of their generally high printing speed.

2.5.3   Hard Faults and Yield

Sufficient manufacturing yield is critical for a flexible TFT technology to allow the integration of complex circuitry and guarantee its commercial viability. Typically, a TFT yield > 99.5 per cent is required for circuits of moderate complexity (i.e., with a transistor count of a few hundreds). A primary yield loss is determined by so-called hard faults,[9] i.e., compromised integrity of the layers in the

---

[9] The associated yield is thus referred to as *hard yield*.

device stack that cause device failure. For instance, hard faults are shorted drain and source terminals due to patterning imperfections, and shorted gate and source/drain terminals due to pinholes in the gate dielectric, or particles/imperfections on the substrate. Hard faults can also appear in interconnecting lines and vias. Organic and metal-oxide TFT technologies are inherently more prone to hard faults than conventional IC manufacturing (e.g., due to the significant defect count of flexible substrates, particle generation via substrate handling or the adoption of solution processing). While organic and metal-oxide TFT processing is often characterised by low hard yield at an academic lab scale (typical achievable transistor count in the order of a few hundreds), transfer to industrial production lines has led to yields close to 100 per cent in unipolar technologies (e.g., via solution-processed organic [102], [103] and vacuum-based metal-oxide TFTs [104]). Due to the more recent development of complementary technologies, their hard yield has been assessed only in a very limited number of cases. In spite of the greater yield challenges of complementary integration, a notable result is the 98 per cent hard yield achieved by the CEA-Liten group on a fully printed organic complementary platform [105].

2.6 Reliability and Variability of Organic and Metal-Oxide TFTs

This section discusses two effects that impact predictability and yield of circuits based on organic and metal-oxide TFTs. These relate either to changes of device performance over time and under prolonged electrical stress (Section 2.6.1) or to statistical variations across a number of nominally identical TFTs (Section 2.6.2). Both effects are of particular importance in flexible analogue electronics, in view of its sensitivity to the detailed shape of the transistor characteristics, and its frequent reliance on the close matching of nominally identical devices.

2.6.1 Bias-Stress Effects

The performance of a TFT may not be fully reproducible after prolonged operation. If its characteristic parameters (e.g., threshold voltage, carrier mobility, subthreshold slope) change over time during operation, the electrical bias is said to stress the TFT.

As performance stability under bias is crucial for the success of a semiconductor technology, extensive research has been conducted on these effects in organic and metal-oxide TFTs. Most studies have attributed bias-stress effects to trapping of charge carriers. A fundamental distinction has been made between intrinsic effects, which are caused by the very nature of the materials in a TFT, and extrinsic ones, which are traced to undesired chemical species introduced into a TFT during its fabrication or from the atmosphere in which it is operated.

Phenomenologically, trapping of a charge carrier consists in its transition from a mobile state into a less mobile one, in which it remains for an extended time. When a charge carrier in the channel of a TFT becomes trapped, it behaves as a fixed charge, and does not contribute to the drain-source current. In a TFT, charge trapping can occur in the semiconductor, in the dielectric, or at their interface, and its most direct consequence is a shift of the threshold voltage in the direction of the applied gate bias.

A number of bias-stress effects have been observed in organic TFTs. The most typical behaviour takes the form of a rigid stretched-exponential shift of the transfer characteristic in the direction of the applied bias with time (see Figure 2.8a), which falls within the picture of charge trapping.

The bias-stress behaviour of metal-oxide TFTs has also been extensively characterised, and the instability of top performance metal-oxide semiconductors has been found to be generally less pronounced than that of organic TFTs. The most frequently reported behaviour is a positive threshold voltage shift when stressed in electron accumulation, with no change in subthreshold slope and mobility [85].

Thanks to the considerable insight acquired over the years, great progress has been made in the reliability of metal-oxide-

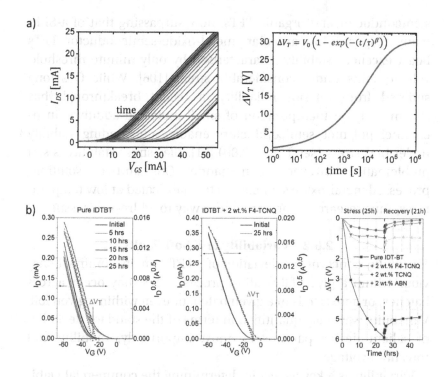

Figure 2.8 a) Illustration of constant-voltage bias-stress degradation in an organic transistor typically observed in the early days of organic electronics: transfer characteristics (left) and threshold voltage shift (right) over time. The degradation here takes the form of a stretched-exponential V_T shift in the direction of the applied gate bias with time. b) Recent breakthrough in the environmental stability of solution-processed polymer TFTs (IDT-BT/CYTOP® stressed at I_{DS} = 2.5 μA): transfer characteristics under stress of pure IDT-BT TFTs (left), of IDT-BT plus small molecule additive (centre), and threshold voltage shift (right) over time [61]. Polymer TFTs incorporating suitable small-molecule additives (e. g., F4-TCNQ, TCNQ and ABN) undergo negligible degradation in comparison to the pristine case. Adapted by permission from Macmillan Publishers Ltd: Nature ([61]). Copyright (2016).

semiconductor and organic TFTs, now surpassing that of a-Si in many respects. In particular, metal-oxide-semiconductor TFTs boast excellent stability, characterised by only minute threshold voltage shifts under considerable stress [106]. While OSs long suffered from inferior reliability, a recent breakthrough has shown that the incorporation of small-molecule additives in p-channel polymer semiconductors enables outstanding stability under bias stress (see Figure 2.8b) [61]. Bias stress stability is still problematic, however, in n-channel OSs and in solution-processed metal-oxide semiconductors fabricated at low temperatures, thus research efforts are under way to address this issue.

2.6.2 Variability and Soft Faults

Variability concerns the deviation of a TFT from its nominal behaviour within a given fabrication process. This may occur across batches, or across foils of a given batch, or even within a given foil. Variability is usually quantified in terms of the standard deviation of salient device parameters, most importantly, mobility and threshold voltage.

Variability is a key aspect in determining the commercial viability of a flexible TFT technology. Indeed, if a TFT deviates substantially from its nominal behaviour, it can compromise circuit functionality. This situation is referred to as *soft fault*, and contributes to determine the soft yield of a given technology. Commercial exploitation requires that a total yield (accounting for both soft and hard faults) compatible with the targeted circuit complexity is achieved.

Variability has long been an important issue in organic semiconductor technologies, especially if based on solution processing. Organic TFT variability may arise, for instance, from structural non-uniformities in the semiconductor film and variability of processing conditions. Reports on organic TFT variability generally focus on intra-foil variations (variations within a foil), giving typical standard deviations in the range of 20 − 50 per cent of the nominal values [107]-[112].

If matched pairs of TFTs are considered, i.e. closely spaced TFTs that are nominally identical in layout, the standard deviation of the parameter difference between the matched devices scales inversely with the square root of the active area (i.e., $W \cdot L$), thus conforming to the same trend observed in silicon technology [108], [109]. In other words, large transistors allow improved matching, at the cost of increased power dissipation and area consumption.

From a material point of view, variability is a lesser problem in near-amorphous polymer systems, as they are not burdened with the microstructural non-uniformities typical of polycrystalline counterparts. Furthermore, it has been recently shown that trap-healing of p-channel polymers via incorporation of small-molecule additives can reduce substantially device spread to a standard deviation much smaller than 10 per cent [8], [61].

AMOxS technologies based on sputter coating generally present limited device spread. This has been attributed to the amorphous nature of these materials, which makes them insensitive to local inhomogeneities [113], [114]. The solution-processed counterparts, instead, present greater variability, especially at low processing temperatures [115], [116].

In summary, organic and metal-oxide TFTs may present variability and mismatch much greater than conventional silicon technology. If this leads to soft faults, variability limits the achievable complexity of transistor integration. Furthermore, even in the absence of soft faults, variability is detrimental to the performance of flexible analogue electronics, as discussed in the following chapters.

3 Flexible Analogue Amplifiers

This chapter covers the fundamentals and recent developments in analogue amplifiers realised with organic and metal-oxide semiconductors. For the sake of simplicity, we refer to them as *flexible analogue amplifiers* or *flexible amplifiers* in the rest of this Element.

Due to the centrality of signal amplification in analogue applications, amplifiers constitute a key circuit block of flexible analogue electronics. This chapter first examines the general concepts pertaining to analogue amplification (Section 3.1). Subsequently, the technologies and circuit architectures specific to flexible amplifiers are discussed (Section 3.2). Finally, the developments and state of the art of flexible amplifiers are presented, combining a discussion on material-related breakthroughs and on the milestones in the main performance parameters (Section 3.3).

3.1   General Concepts

3.1.1   Transfer Characteristics, Gain, Bandwidth, Noise and Sensitivity

Signal amplification involves boosting the amplitude of a signal to ensure the reliability of its subsequent processing. This function is made necessary to prevent the signal of interest from being corrupted by added electronic noise and interference. Amplification is quantitatively described in terms of amplification gain, or simply gain, defined as the ratio between the change of the amplifier's output v_O with respect to its input v_I as pictorially represented in Figure 3.1a:

$$A_v := \frac{dv_O}{dv_I}\Big|_{at\ bias\ point} = \frac{v_o}{v_i}.$$

Here v_I and v_O represent the total input and output quantities, while v_o and v_i represent the input and output small signals superimposed on the given operating point, the so-called DC bias point. Most frequently, the input and output signals are time-varying voltages, a case referred to as *voltage amplification*, the associated gain being in units of V/V. In consideration of its widespread adoption, we tacitly refer to voltage amplification in the remainder of this chapter.

The derivative in the definition of gain makes it apparent that amplification refers to small variations of the amplifier's input and output. Underlying this is the fact that amplifiers are realised with nonlinear devices (i.e., transistors), hence only small perturbations of their terminal voltages and currents ensure that the relationship between input and output signals is linear to a good approximation.

Although amplification is a small-signal concept, the evaluation of an amplifier's performance also requires a detailed understanding of its large-signal behaviour, i.e., its response to large variations of the input signal. The functional dependence of the amplifier's output voltage with respect to its input over the full rail-to-rail range (defined by the power supplies) constitutes the amplifier's voltage transfer characteristic (VTC). An example of amplifier VTC is given in Figure 3.1c, and it shows that amplification is possible only within specific input and output signal ranges. For too large an input signal, the amplifier's component transistors may exit the region of operation suitable for amplification. Likewise, the output signal may be clipped as it approaches the power rails. Ideally, it is desirable that symmetric variations of the input signal around the bias point are possible while maintaining the nominal gain value, with the output terminal being capable of spanning the full rail-to-rail range.

In addition to increasing the signal amplitude, an amplifier is expected to maintain the detailed temporal evolution of the signal of interest, as that constitutes an essential element of the signal's analogue character. This feature is determined to a significant extent by the maximum signal frequency that can be amplified by $\approx A_v$, given that the gain typically drops off at high frequencies (Figure 3.1d). In particular, special significance is associated with the frequency f_{3dB} at which the gain is reduced by a factor of $\sqrt{2}$ (Figure 3.1d), which defines the amplifier bandwidth. An additional characteristic frequency is the frequency *GBW* at which the gain reaches unity (i.e., 0 dB), which is referred to as *unity-gain frequency* or *gain-bandwidth product*.

While boosting the signal of interest to prevent its corruption by noise added by the signal-processing chain, an amplifier also introduces some (ideally negligible) noise of its own. This is

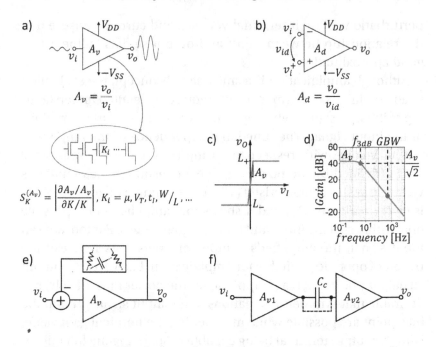

Figure 3.1 Fundamental configurations, parameters and characteristics of analogue amplifiers. a) Single-input amplifier, voltage gain and its sensitivity to variations of component transistors' parameters. b) Differential amplifier and differential gain. c) Voltage transfer characteristic of a differential amplifier, which has a linear response only over a narrow differential input voltage range. d) Frequency response of an amplifier, illustrated in terms of the magnitude of the voltage gain as a function of input signal frequency. e) Feedback amplifiers comprising a feedback network generically composed of resistors and capacitors. f) Example of multistage amplifier, featuring two amplification stages connected either by direct coupling (dashed line) or by a coupling capacitor C_c.

typically modelled through equivalent noise sources at its input terminals. The amplifier input noise adds to the noise inherent in the signal to be amplified. The total resulting noise thus sets the input-referred signal-to-noise ratio *SNR*:

$$SNR = \frac{P_s}{P_{n,s} + P_{n,a}}.$$

Here P_s is the input signal power, and $P_{n,s}$ and $P_{n,a}$ are the noise power associated with the input signal and the equivalent input noise power of the amplifier.

Variations of process parameters inevitably cause the amplifier to deviate from its nominal behaviour. Denoting with Q any of the characteristic amplifier parameters (e.g., bias point, gain, bandwidth), and with K a specific process parameter (e.g., carrier mobility, threshold voltage, gate dielectric capacitance), it is customary to describe how Q is impacted by process variations of K in terms of its sensitivity, defined as

$$S_K^{(Q)} := \frac{\partial Q/Q}{\partial K/K}.$$

The smaller the sensitivity, the more a technology and/or circuit topology is robust with respect to process variations.

3.1.2 Single-Input versus Differential Amplifiers

In the simplest arrangement possible, amplification is carried out via single-input amplifiers (Figure 3.1a). As the name suggests, one such amplifier is equipped with only one input terminal, and the input and output signals are intended as referenced to the circuit ground. While advantageous for their simplicity, single-input amplifiers bear significant limitations. Firstly, they are more prone to interference, which may be particularly problematic in sensing applications. Indeed, the signal from a sensor often comes across two leads, which may pick up some common interference. If the two leads are connected to a single-input amplifier, the sensor's signal and the interference will be amplified indiscriminately, and signal integrity may be compromised. Additionally, extra care must be taken when feeding a signal to a single-input amplifier, and when connecting a number of such amplifiers in a multistage configuration (see Section 3.1.4). In fact, direct coupling (i.e., the direct connection between the two relevant terminals) is generally not possible in either case, because it would perturb the amplifier's bias point and compromise its functionality. Coupling capacitors are often required to circumvent the problem (Figure 3.1f). These may pose an additional technological challenge and can be detrimental to the application at hand (e. g., when low-frequency signals are to be amplified).

In contrast to single-input amplifiers, differential amplifiers are equipped with two input terminals, across which the input signal is applied (Figure 3.1b). Although they generally involve a higher transistor count than single-input counterparts (typically by about a factor of two), differential amplifiers are superior in terms of signal integrity. Indeed, a differential amplifier boosts the difference signal v_{id} between its two input terminals ($v_{id} := v_i^+ - v_i^-$, where +/– denote the two input terminals) by the differential gain $A_d := v_o/v_{id}$ (Figure 3.1b). At the same time, components common to both input terminals ($v_{icm} := (v_i^+ + v_i^-)/2$) are amplified by a much smaller (ideally zero) gain $A_{cm} := v_o/v_{icm}$, referred to as *common-mode gain*. Hence, the ratio A_d/A_{cm} (the so-called common-mode rejection ratio, *CMRR*) is usually much greater than unity (ideally infinite). This means that interference appearing at both input terminals is suppressed. In virtue of this, differential amplifiers provide greater signal integrity than single-input counterparts. Therefore, differential amplifiers often constitute the circuits of choice for analogue amplification, and this is especially true for sensing applications.

3.1.3   Feedback Amplifiers

Analogue amplification is often realised by using an amplifier within a negative feedback circuit, namely, a (negative) feedback amplifier. In this case, the amplifier serves as a gain block, and the negative feedback consists of a (usually passive) network connected between the amplifier's input and output terminals via a summing node (Figure 3.1e). While typically leading to much smaller gains than in absence of negative feedback, feedback amplifiers enable a series of benefits for signal amplification. Most importantly, feedback amplifiers show reduced sensitivity to process variations, broader bandwidth and an enhanced response linearity. These advantages have determined their ubiquitous adoption in analogue electronics.

For an amplifier to be used as a gain block in a feedback amplifier, it must meet specific requirements. To give an example, for negative feedback to deliver its intended benefits, the magnitude of the amplifier gain has to be sufficiently large. Indeed, general-purpose silicon-based amplifiers (so-called operational amplifiers) employed in conventional negative feedback circuits have a typical gain larger than 10^5 V/V.

3.1.4 Multistage Amplifiers

Multistage amplifiers are a frequent solution when high gain figures are needed, e.g., for the gain block of a feedback amplifier. Multistage amplifiers involve connecting the output of one amplifier (referred to as *amplification stage* in this context) to the input of another (Figure 3.1f). This approach delivers an overall gain equal to the product of the gains of the component stages, provided that no significant loading occurs (i.e., the input resistance of each of the stages is sufficiently large).

Multistage amplifiers are widely adopted in silicon technology, where a typical configuration consists of a first differential gain stage, a second single-input gain stage and, finally, a unity gain stage with substantial current-driving capability. An even larger number of stages have often been used in flexible analogue electronics, especially when the specific technology or circuit topology would otherwise lead to low gain figures [117].

An important requirement for multistage amplifiers is that the bias point of any given stage falls within the input range of the subsequent stage. This indeed enables direct coupling. Additionally, if used in a negative feedback configuration, a multistage amplifier should be stable even if connected in unity-gain feedback. This requirement can be cumbersome to achieve,

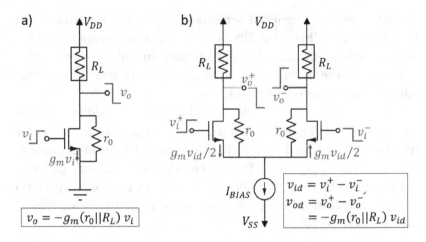

Figure 3.2 Operation of basic amplifier configurations: a) single-input (common-source) amplifier and b) differential amplifier. The output resistance r_0 is shown explicitly for the sake of convenience. The rectangles around the load resistors are meant to indicate that the load devices can be resistors or devices with equivalent resistant R_L.

especially if the number of gain stages in the multistage amplifier is greater than two.

3.1.5 Transistors as Building Blocks for Analogue Amplifiers

The simplest amplifier consists of a transistor biased in saturation, the input signal being applied between its gate and source terminals, the source being connected to signal ground and the drain to the power supply via a load device of resistance R_L (actual or equivalent).[10] A pictorial representation is given in Figure 3.2a. This configuration, called *common source*, allows the transistor to behave as a variable signal current source controlled by the input

---

[10] While other configurations using a single TFT are possible, here we focus on the common-source one as it is relevant to voltage amplification, in line with the main focus of this chapter.

signal at its gate: $i_d = g_m v_i$. The resulting current signal is fed to the load device to convert it to a voltage signal: $v_o = -g_m(R_L \parallel r_0)v_i$, where r_0 is the transistor output resistance (see Chapter 2). The voltage gain here is given by the input transistor's transconductance multiplied by the cumulative load at its drain terminal: $A_v = -g_m(R_L \parallel r_0)$. It must be emphasised that the load device not only allows the current-to-voltage conversion but is also essential for the correct biasing of the input transistor, as it concurs in setting its DC drain voltage. R_L can be a resistor, in which case the quiescent drain voltage is set to $V_{DD} - R_L I_D$. In this case, selecting extremely large resistance values – an attractive solution to maximise the amplifier's gain – would not be viable if the voltage drop across the load resistor pushes the input transistor out of the saturation region. This limitation can be overcome, however, if the load device is realised via a transistor. The non-linearity of the load transistor's $I_{DS} - V_{DS}$ characteristics can then be exploited. This typically affords a large equivalent load resistance with a limited voltage drop across. In this latter case, the load is denoted as *active*, in contrast to the passive (i.e., resistive) load realised via a resistor, and the resulting amplifier architecture is called active-loaded.

The simplest differential amplifier is an immediate extension of the single-input amplifier above. It consists of two branches, each one containing an input transistor and a load device (Figure 3.2b). The two input transistors have their source terminals connected to the same current source (the so-called tail current source), and they are nominally identical, just as the load devices. When a differential input voltage is applied across the input transistors' gate terminals, the resulting current signal circulates around the two branches of the amplifier, leading to a differential output voltage equal to $v_{od} = -g_m(R_L \parallel r_0)v_{id}$. When a common-mode input signal is applied, instead, the voltage signals across the two load devices change by the same amount. This leads to a zero differential output voltage, and corresponds to the ideal case of infinite *CMRR*. The *CMRR* in real differential amplifiers may be large but is not infinite, due to device parameter mismatch between the two branches of the differential pair and because of the finite output resistance of the tail current source.

3.2   Circuit Architectures

Building on the fundamental concepts discussed in Section 3.1, we now examine the specifics of flexible amplifiers in terms of circuit architectures. As will become apparent, a recurrent theme is the distinction between unipolar and complementary flexible technologies, which are thus introduced first. This section serves as the foundation for the subsequent discussion on the developments and state of the art of flexible amplifiers presented in Section 3.3.

3.2.1   Unipolar versus Complementary Technologies

In the realm of flexible amplifiers, a basic distinction between technologies concerns the channel polarity (i.e., p- or n-channel) of their component transistors. Unipolar amplifiers involve transistors of one single channel polarity, while complementary ones comprise devices of both. Incidentally, organic TFTs with ambipolar conductivity are also available, yet they have not been employed for analogue amplification thus far due to inherent difficulties in biasing, and resulting design complexity. This fundamental technological distinction applicable to flexible analogue amplifiers has no direct equivalent in silicon-based analogue electronics, where amplification at the state of the art is always carried out via complementary circuits.

Flexible unipolar amplifiers have been pursued more frequently than complementary ones. The reason is twofold: on one hand, limitations of performance and/or stability inherent in organic and metal-oxide semiconductors; on the other, processing complexity. From a material point of view, as discussed in Chapter 2, OSs count a number of high-performance p-channel materials, while the n-channel counterparts are generally inferior; conversely, metal oxides are excellent as n-channel semiconductors, while the search for p-channel metal oxides with matching performance is still ongoing. Incidentally, performance is to be intended here in terms of transistor mobility and threshold voltage, the values of which should be closely matching in both n-channel and p-channel devices for complementary integration to be viable.

In terms of processing, it is evident that a unipolar approach does not require in principle any patterning of the semiconductor film when the intrinsic conductivity of organic or metal-oxide semiconductors is low enough to ensure isolation between adjacent devices. Therefore, unipolar technologies can resort to convenient blanket deposition methods, which streamline circuit fabrication and allow better control of device parameters. Complementary technologies, instead, do require patterning of the n-channel and p-channel semiconductors. A critical requirement here is that the electronic properties of the semiconductor that is deposited first are not compromised by the deposition and patterning methods of the semiconductor laid down last. For instance, this may prevent processing steps that would normally be used to achieve optimal performance with stand-alone TFTs (e.g., surface treatments via self-assembled monolayers, oxygen plasma or UV-ozone), and poses important restrictions on the solvents that can be used in the ink formulation of the semiconductor deposited last. It follows that complementary integration is particularly challenging from a processing point of view.

3.2.2 Basic Loading Strategies

Most flexible amplifier implementations explored thus far feature an active-loaded architecture. This is not only in view of circuit considerations (see discussion in Section 3.1.5), but also because of the challenge of integrating purely resistive loads in an organic or metal-oxide technology. In fact, organic and oxide semiconductors generally come in an intrinsic form to realise enhancement-mode transistor circuits, and thus possess an extremely high intrinsic resistivity. This implies that the realisation of resistive loads with resistance in the desired range would require one or more additional processing steps – the selective doping of the semiconductor in the load device or the deposition and patterning of another

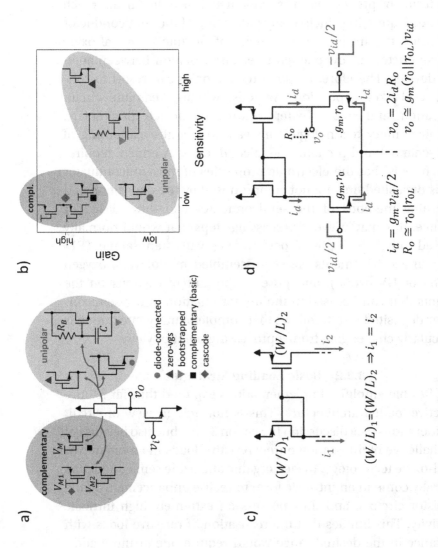

Figure 3.3 Single-input amplifiers in complementary and unipolar technologies: a) loading strategies, and b) corresponding relation between gain and sensitivity. Common loading strategy in differential complementary amplifiers: c) operation of a current mirror, and d) example of a current-mirror-load differential amplifier.

semiconductor inclusive of some dopant. Given the increase in complexity, this option is of limited practical appeal.

Active-loaded flexible amplifier implementations are predominant in both unipolar and complementary technologies. As a matter of fact, from a circuit design point of view, the two differ precisely in the type of active load employed: while complementary technologies allow the loading with a transistor of opposite channel polarity, several options are available for unipolar implementations – for instance, see Figure 3.3a. Here we present the most basic loading strategies adopted in flexible analogue amplifiers, first discussing the ones relevant to single-input amplifiers and subsequently covering the differential counterpart. This discussion makes reference to the parameters defined in Figure 3.3. More complex loading strategies and amplifier architectures are examined in Section 3.2.3.

Single-Input Amplifiers

A possibility relevant to unipolar technologies is to load a single-input amplifier with a transistor biased with zero gate-source voltage, a configuration known as *zero-vgs load* (Figure 3.3a). This approach has received some attention in unipolar technologies as a way to achieve large load resistance and gain [118]. A transistor biased at $V_{GS} = 0\,V$ is normally operating in its subthreshold region, hence its output resistance is extremely high. This allows the amplifier gain A_{0VGS} to approach the intrinsic gain of the input transistor: $A_{0VGS} \approx -g_m\,r_0$. At the same time, the amplifier bias point is set by the current of the load transistor at $V_{GS} = 0\,V$. As the load TFT is operated in the subthreshold region, the bias point is thus a very strong function of the threshold voltage. This implies that small variations of process parameters will determine a dramatic change of the amplifier bias point. A generally poor control of the amplifier bias point and gain is likely to follow, therefore this configuration is inherently characterised by high sensitivity to process parameter variations (Figure 3.3b). Finally, note that a zero-vgs load can be used only if the load TFT is still conductive at $V_{GS} = 0\,V$ (normally-*on* TFT).

A more widely explored strategy in the flexible amplifier literature consists in having a diode-connected transistor as a load
(Figure 3.3a) [2], [119]–[123]. In this case, the drain and gate of
the load transistor are shorted. The equivalent load resistance of
this configuration is approximately equal to $1/g_{mL}$, where g_{mL} is the
transconductance of the load transistor. The gain A_{0VGD} of a diode-
connected-load amplifier is thus fixed by a transconductance
ratio: $A_{0VGD} \approx - g_m/g_{mL}$. Therefore, to a first approximation, the
gain is determined by the transistors' aspect ratios:
$A_{0VGD} \approx - \sqrt{(W/L)/(W/L)_L}$, where $(W/L)_L$ is the aspect ratio of
the load transistor.[11] It follows that the gain of a diode-connected-
load amplifier is primarily determined by transistor geometry.
Moreover, the bias point is set by the current balance of two
transistors both operating in the saturation region with a relatively
large overdrive, thus it is less sensitive to variations of process
parameters (Figure 3.3b).

As is apparent, diode-connected loads offer several benefits,
which have determined their popularity in unipolar amplifiers. At
the same time, they lead to limited gain. While it is true that the
larger the aspect ratio of the input transistor the greater the gain, for
too large an asymmetry between input and load transistors the
voltage transfer characteristic becomes skewed (e.g., the VTC transition can no longer be made to occur around $V_{DD}/2$). Consequently,
multistage amplifiers are not viable (unless coupling capacitors are
introduced). This is besides the purely geometrical consideration
that achieving a large gain would require a very large input transistor, detrimental in terms of circuit area, fabrication yield and speed
(due to parasitic capacitances). This is the reason why literature
implementations adopting this loading strategy have typical gain
values of about ≈ 3 V/V per stage [2], [119]–[122].

In basic complementary amplifier topologies, the drain of the
input transistor is connected to the drain of a load transistor of
opposite channel polarity and biased at a fixed V_{GS}. This leads to a

---

[11] Here we assume that mobility is not a function of the applied gate-source
voltage.

gain $A_c = -g_m(r_0 \parallel r_{0L})$ (where r_{0L} is the output resistance of the load transistor), thus typically of the same order of the intrinsic gain of the input transistor. Furthermore, the bias point is set by the current balance of two transistors both biased in saturation with a relatively large overdrive, enabling rather limited sensitivity to process parameter variations (Figure 3.3b). For reasonably well matched mobility and threshold voltage values, symmetric VTCs are achieved, which generally allow direct-coupled multistage amplifiers [7], [8], [124], [125].

Differential Amplifiers

The discussion on single-input amplifiers is immediately applicable to differential amplifiers. In the simplest implementation of a differential amplifier, the load device is simply replicated on the two branches of the differential circuit, as in Figure 3.2b. Discussion on performance and sensitivity would follow the same arguments examined in the case of single-input amplifiers, with the notable addition that the rejection of common-mode signals in a differential amplifier relies on the close matching of the load devices on its two branches.

An additional basic loading strategy yet to be presented concerns complementary differential amplifiers. Most commonly, complementary differential amplifiers are loaded with a current mirror, i.e., a subcircuit consisting of two transistors connected as in Figure 3.3c [7], [8], [124]–[128]. As its name suggests, a current mirror enables the same current at the drains of its two component transistors, provided that they are of identical aspect ratio. When employed as the load of a differential complementary amplifier, a current mirror enables, by construction, the same signal current in the two branches of the amplifier, as shown in Figure 3.3d. As is apparent, this arrangement leads to superior robustness to process parameter variations, as long as matching is reasonable. Moreover, in terms of gain, a current-mirror load leads to a current at the amplifier's output terminal twice as large as in the basic differential amplifier of Figure 3.2b. Indicating with r_{0L} the output resistance of the transistors in the current mirror, this allows a single-output gain (i.e., the gain at a single output terminal) equal to $A_d = -g_m(r_0 \parallel r_{0L})$. This approaches the intrinsic gain of the input transistors, and is twice as large as the single-output gain of the basic differential amplifier. Therefore, it comes as no surprise that this configuration has led to the highest gain figures reported for flexible analogue amplifiers [7], [8], [124] (Section 3.3).

3.2.3   More Complex Architectures

While many flexible amplifiers investigated to date employ the basic loading strategies discussed earlier, researchers have also developed more complex[12] architectures to achieve superior performance in terms of gain and/or robustness to process parameter variations. These architectures are examined in the rest of this section. The discussion firstly focuses on single-input amplifiers, then deals with differential unipolar amplifiers and finally covers differential complementary amplifiers.

Unipolar Single-Input Amplifiers

In an effort to limit the sensitivity of a zero-vgs-load organic amplifier, Lee et al. demonstrated the applicability of a dual-gate device structure to shift the threshold voltage of the input transistor and, concurrently, the amplifier VTC [129]. This work thus succeeded in compensating the threshold voltage variability at the cost of an extra bias voltage (as needed to control the input transistor backgate). In particular, the tuning of the amplifier VTC via the transistor backgate enabled greater matching of the input-output DC levels.

Addressing the skewedness and limited gain of diode-connected-load amplifiers, Huang et al. pursued a pseudo-CMOS approach [130]. This led to more symmetric VTCs and enabled the post-fabrication compensation of process variations and device degradation. The superior performance thus attained, however, involved a double transistor count and an additional voltage supply.

---

[12] *Complexity* refers here to the higher transistor count and/or the more involved biasing scheme required by these architectures compared to the basic ones of Section 3.2.2.

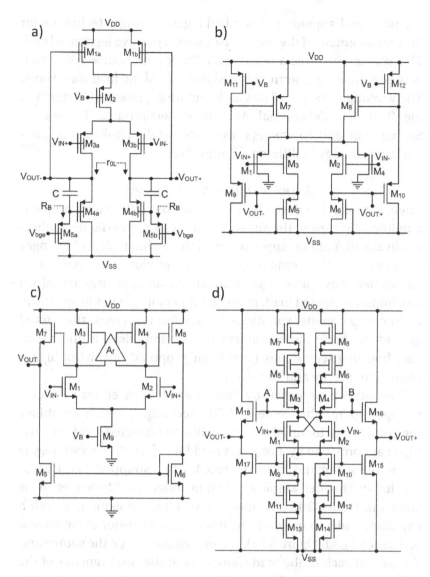

Figure 3.4 Gain-boosting techniques in single-stage unipolar differential amplifiers. a) BGE technique. b) Positive-cum-negative feedback. c) Positive feedback between output nodes and gates of load transistors. d) Positive feedback and enhanced source-degenerated loads in dual-gate unipolar technology.

Lastly, Shabanpour et al. applied a gain-boosting technique for the enhancement of the load resistance via positive feedback [131]. This scheme involves a positive feedback path with ≈ 1 V/V gain between the output terminals and the gate of the load transistors. The working principle of this architecture is presented in detail in the 'Unipolar Differential Amplifiers' section that follows, as Shabanpour and co-workers also successfully applied this gain-boosting strategy to differential amplifiers.

Unipolar Differential Amplifiers

Intense research efforts have focused on differential unipolar amplifiers, in view of the superior immunity to interference inherent in the differential approach, and the technological challenges associated with complementary integration. In particular, researchers have investigated advanced architectures based on the diode-connected load, preferred due to its greater insensitivity to process parameter variations. In an effort to increase the limited gain of the basic diode-connected-load amplifiers, a number of gain-boosting techniques have been proposed, involving higher circuit complexity and transistor count.

The solution proposed by Marien et al. is based on the bootstrapped gain-enhancement (BGE) technique, which combines the robustness to variability of a diode-connected load and the high gain provided by a zero-vgs load [3]–[5], [117]. This strategy is illustrated schematically in Figure 3.3a ('bootstrapped' load), while the full p-channel amplifier demonstrated by Marien et al. is shown in Figure 3.4a. A high-pass RC filter, implemented with a capacitor and a transistor (which serves as a resistor of equivalent resistance R_B, cf. Figure 3.3a), is connected between the source and the gate of each of the load transistors in the two branches of the differential amplifier. The amplifier operation can be analysed in DC and for a sinusoidal small signal. In DC the transistors $M_{5a\text{-}b}$ discharge the gates of the load transistors $M_{4a\text{-}b}$ to V_{SS}. The load transistors are thus diode-connected. This enables reduced sensitivity of the bias point just as for a diode-connected load. For a sinusoidal signal, at a frequency greater than $f_c = 1/(2\pi R_B C)$, the impedance offered by the capacitors C is much smaller than the impedance offered by the transistors M_{5a-b}. The gates of transistors M_{4a-b} are then shorted to their respective sources, and the load transistors behave as zero-vgs devices, offering an output resistance r_0. This entails that at frequencies above f_c the current signal from each input transistor flows into the parallel between a bootstrap resistor R_B and the load output resistance r_{0L}, thereby achieving a stage gain $\cong - g_m(R_B \parallel r_{0L})$, where g_m is the input transistors' transconductance and r_{0L} is the parallel of the output resistance of input and load TFTs. This generally allows an improvement of the gain with respect to the basic diode-connected-load configuration (R_B can be tuned to values $\gg 1/g_{m4a-b}$). Inevitably, this arrangement comes with the drawback that signal components below f_c are amplified much less. Moreover, the relatively larger gain is achieved at the price of greater complexity and area. At last, we note that the circuit demonstrated by Marien et al. also includes the following features, inessential for the BGE technique yet beneficial for circuit performance: a) a gate-backgate steering technique (based on double-gate transistors), employed to increase the transconductance of the input transistors; b) a common-mode feedback loop to reduce the sensitivity of the output bias point to TFT variability.

An alternative gain-boosting technique was proposed by Chang et al. and is based on a positive-cum-negative feedback concept (Figure 3.4b) [120]. Transistors M_1-M_2 and M_3-M_4 form two input differential pairs, biased with the tail transistors M_7 and M_8 and loaded with diode-connected transistors M_5 and M_6. The output of each differential pair (source terminals of M_5 and M_6) is fed back to the tail transistors (M_7 and M_8) of the other pair through a level shifter (M_9-M_{11} and M_{10}-M_{12}). This establishes a positive feedback loop that enhances the amplifier's gain: the tail current sources enhance the current signal through the branches of the differential pair in response to variations of the output terminals, thus leading to a greater differential output signal. Besides this gain-boosting scheme, the design of Chang et al. incorporates a negative feedback loop to reduce the sensitivity of the output common-mode level. This can be illustrated by assuming that, for instance, due to process parameter variations (e.g., threshold voltage variations), the output common-mode voltage rises. This will lead to a reduction of the overdrive voltage of M_7 and M_8, and, consequently, to a reduction of the bias currents of the differential pairs. In turn, a reduction of V_{OUT+} and V_{OUT-} will follow, thus counteracting the initial rise in the output common-mode level.

A different gain-boosting scheme was applied by Shabanpour et al. [132] to a metal-oxide differential amplifier. Derived from a single-input amplifier architecture investigated earlier on [131], the underlying idea is to control dynamically the overdrive voltage of the load transistors, in such a way that the transistors resemble an active load. If an amplifier A_f is used in a feedback configuration to regulate the overdrive voltage of the load transistors (Figure 3.4c), the expression of the equivalent small-signal load resistance at the drain of M_1 reads:

$$R_L = r_{01} \| \frac{1}{g_{m3}(1 - A_f)}. \tag{3.1}$$

For $A_f \to 1^-$, the effective load resistance becomes $\approx r_{01}$, thus leading to a voltage gain close to the intrinsic gain of the technology. This is almost equivalent to a zero-vgs load for small signals, while the load's DC overdrive remains similar to that in a diode-connected load, thus taking advantage of the features of both load types. It is worth noting that if $A_f > 1$, the load provides an equivalent negative resistance which makes the amplifier unstable. Thus a feedback gain slightly smaller than unity must be adopted in order to ensure stability even in the presence of process variations.

As in single-ended amplifiers, the performance of differential amplifiers can be improved using a dual-gate device structure. Novel single-stage differential amplifier architectures, based on dual-gate metal-oxide TFTs, were demonstrated by Garripoli et al. [133]. Significant improvement of DC gain was achieved via positive feedback and enhanced-source degenerated loads. The gain enhancement due to the positive feedback was initially demonstrated on a simple differential amplifier with diode-connected loads. The second gate of the driver transistor in each branch is connected to the output of the opposite branch, creating a positive feedback loop (Figure 3.4d).

Additionally, in order to increase the output resistance and thus the DC gain, Garripoli et al. replaced the standard diode-connected load with a new load topology realised with a stack of three diode-connected transistors (M_7-M_5-M_3 and M_8-M_6-M_4) with a peculiar interconnection of the second gate electrodes (Figure 3.4d) [133]. The load output resistance at node A (a similar argument applies to node B) is dependent on the threshold modulation coefficient η (see Chapter 2), and its expression reads:

$$R_{OUT} = \frac{1}{g_{m7}} + \frac{1}{g_{m5}}(1+\eta) + \frac{1}{g_{m3}}(1+\eta)^2.$$

Since η is bigger than 1 for the technology used in this work, the total output resistance is enhanced, and thus the DC gain.

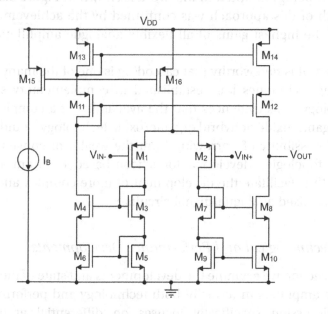

Figure 3.5 Complementary cascode stacked-mirror amplifier realised by Maiellaro et al. [7].

Complementary Differential Amplifiers

To achieve higher gain figures, a possible improvement of the basic complementary differential amplifier involves the enhancement of its output resistance. A classical approach is cascoding, which involves connecting the drain of each load transistor to the respective amplifier's output node via a cascode transistor (cf. Figure 3.3a). This leads to the enhancement of the apparent output resistance by a factor equal to the intrinsic gain of the cascode transistor, in turn resulting in a higher voltage gain. The only drawback of this method is that the extra transistor in the path between power rails reduces the output common-mode range. In other words, a cascode load requires a higher power supply voltage to achieve the same output common-mode range of the basic complementary architecture.

Cascoding was successfully applied to organic complementary differential amplifiers by Maiellaro et al. [7], and the specific circuit topology adopted in this work is shown in Figure 3.5. The strength of this approach was confirmed by the achievement of one of the highest gains of all flexible analogue amplifiers (i.e., 49 dB).

Finally, it is noteworthy that cascoding is one of the many gain-boosting techniques long established in complementary silicon technology. This evidences that the availability of a complementary organic, oxide or hybrid organic/oxide technology would open up the possibility of borrowing from the wealth of conventional circuit topologies developed for silicon-based electronics, and would thus facilitate the development of more complex and performant mixed analogue/digital circuits.

3.3   *Technological and Performance Developments*

In this section we examine the developments and state of the art of flexible amplifiers in terms of both technology and performance. The discussion specifically focuses on differential amplifiers (which will be simply referred to as *amplifiers*, for the sake of conciseness), as they constitute the most fitting indicator of the

development of an analogue amplifier technology. Indeed, they are
the prototypical amplifier that possesses a clear and exclusive
analogue character, and importantly so in a research field domi-
nated by digital works such as flexible electronics. In contrast,
single-input amplifiers greatly overlap in architecture and characteristics with the simplest digital gate of all, the inverter.[13]
Secondly, differential amplifiers are superior in several respects,
such as common-mode rejection and signal integrity. These properties are of great interest for sensing applications, a key target area
of flexible analogue electronics.

3.3.1 A Materials Perspective

Here we trace the development of analogue amplifiers from the
point of view of material technologies, first examining the choice of
semiconductors and then dealing with gate dielectrics.

Semiconductors

Although analogue amplifiers based on OSs were reported as early
as in year 2000 (Kane et al. [1], [134]), analogue amplification did
not attract much attention from the organic semiconductor community for the whole subsequent decade. The only other studies
reported in the intervening years were conducted by Gay et al. [2],
[119]. This is in contrast to the more vibrant scene of digital organic
electronics, which concurrently saw formidable demonstrations in
the form of fully printed circuitry [135], and large integration [136],
[137]. This contrast is not simply due to the general preponderance
of digital electronics. In fact, it is related to the inherent properties
and limitations of OSs: analogue circuits are dependent on the
detailed shape of the component transistors' characteristics, and
are more prone to failure in the presence of substantial process
variations, such as is the case of organic and, more generally,
solution-processed semiconductors. Indeed, it is revealing that

---

[13] The most classic example of overlap is the case of unipolar diode-connected-
load or zero-vgs-load single-input amplifiers, which for large input signals can
be regarded as digital inverters.

the early works led by Kane and Gay utilised vacuum-sublimed pentacene, vacuum deposition allowing better control of the electronic properties. Subsequently, amplifiers based on vacuum-sublimed pentacene were also realised by Nausieda et al. [118].

The very first amplifiers with solution-processed OSs were reported by Marien et al. in 2010 [3], [5], [117], and employed solution-processed precursor-based pentacene. These works were also the first to combine a flexible substrate and a solution-processed dielectric, consistently with the technological appeal of solution-based processing. A further step in this direction was taken by Chang and co-workers, who developed a fully additive solution-based process combining TIPS-pentacene and a polymer resin as semiconductor and gate dielectric, respectively [120].

All demonstrations presented so far in this section, spanning more than a decade (2000–2014), are pentacene-based and, most notably, unipolar. This reflects the difficulty of developing an organic-based complementary analogue technology, due to the generally inferior performance and stability of n-channel OSs. The first report of complementary organic amplifiers is that of Vadiya and co-workers and dates back to 2011 [126]. Their amplifiers employed small-molecule p- and n-channel semiconductors (pentacene and C_{60}, respectively), which were both deposited by vacuum sublimation, consistently with the greater control of the transistor characteristics allowed by vacuum-based methods. Another significant complementary analogue technology platform was developed a year later by Ishida et al., and is based on vacuum-sublimed DNTT and NTCDI as p- and n-channel semiconductors [128]. This departure from traditional organic semiconductor systems (i.e., pentacene and C_{60}) reflects the concurrent synthetic developments towards higher mobility and more stable OSs. This is especially true for n-channel materials, which had witnessed the emergence of rylene-based diimides.

A major step towards high-performance amplifiers was the development of a fully additive complementary technology by the CEA-Liten group. Their work was originally based on a PTAA derivative and an acene-based diimide as p- and n-channel semi-conductors, both deposited from solution by screen printing on a plastic substrate [127], [138]. Subsequently, this technology was improved by substituting PTAA with much higher-mobility TIPS-pentacene, thus achieving a more balanced performance of the complementary semiconductor pair [110], [139]. The robustness of this technology enabled high-performance complementary amplifiers, as can be found in the work of Guerin et al. [127], Maiellaro et al. [7], [124] and Abdinia et al. [125].

Besides organic technologies, the flexible electronics community has also investigated analogue platforms using metal-oxide semiconductors [6], [122], [123], [132], [133], [140]. This much more recent development (the first report on the subject was produced by Tai et al. in 2012 [6]) is due to the later emergence of metal-oxide semiconductors. All metal-oxide-semiconductor amplifier reports share a similar approach in terms of semiconductor processing, as they all utilise n-channel indium gallium zinc oxide (IGZO) deposited by sputter coating. This unipolar n-channel approach is reflective of the lower performance of p-channel oxide semiconductors. Furthermore, the use of sputter coating (instead of solution processing) as the semiconductor deposition technique of choice can be understood in relation to the unipolar approach. Indeed, unipolar metal-oxide technologies are compatible with blanket-deposited semiconductor films, which are most conveniently produced by sputter coating.

In view of the complementary strengths of organic and metal-oxide semiconductors, Pecunia et al. recently developed a hybrid approach to solution-processed complementary amplifiers [8]. This work brings together a high-mobility p-channel polymer (IDT-BT) and an n-channel metal-oxide semiconductor (indium zinc oxide (IZO)), both processed from solution on an ultrathin plastic foil. This approach moves away from small-molecule organic semiconductor systems, in view of the constraints posed by the latter on the choice of gate dielectric (see following discussion on dielectrics). A novel trap-healing technique on the polymer semiconductor and the compatibility of the hybrid semiconductor pair enabled amplifiers with very high performance while operating at the lowest power supply voltage to date.

Dielectrics

A number of insulating materials and processing techniques thereof have been explored over the years with the aim of achieving good semiconductor-dielectric interfacial properties within an analogue circuit integration platform.

In the context of organic technologies, it is noteworthy that most of the early amplifiers relied on vacuum-deposited gate dielectrics (sputtered SiO_x, ALD alumina, parylene-C via chemical vapour deposition (CVD)), especially in combination with vacuum-sublimed semiconductors [1], [118], [121], [126], [128], [134]. This can be easily understood by considering the greater control allowed by vacuum-deposited dielectrics on their interfaces with OSs.

As for solution-processed dielectrics in organic amplifiers, P4VP was the first polymer to be evaluated (2006–2007) [2], [119], and soon after it found widespread use due to its robustness [3]–[5], [117]. Solution-based low-κ fluoropolymer gate dielectrics gained momentum later on (2011–2013) [110], [127], [138], [139], and featured in a large number of subsequent amplifier reports [7], [124], [125], [127]. Low-κ fluoropolymers are renowned for the excellent interfaces (characterised by a very low trap density [141], [142]) they form with OSs, and were thus essential for the high performance of all these integration platforms. Furthermore, the benign properties of low-κ fluoropolymers also manifest at a processing level: within the CEA-Liten platforms relying on a TG transistor architecture [7], [124], [125], [138], fluoropolymers are a unique option in that they can be deposited from solvents that do not attack the underlying small-molecule semiconductors.

While remaining in the realm of solution-processed dielectrics, the hybrid organic/metal-oxide platform of Pecunia et al. [8] took one step further by utilising a bilayer dielectric, which comprises a low-κ fluoropolymer as an ultrathin (25 nm) interfacial layer, and a relaxor ferroelectric terpolymer as the bulk of the dielectric stack [8]. This solution was allowed by the departure from small-molecule semiconductors (dominant until then in analogue organic platforms), which relaxes the solvent-related constraint on the choice of gate dielectric in a TG transistor technology.[14] In addition to exploiting the excellent organic semiconductor–dielectric interface allowed by the low-κ fluoropolymer dielectric, this solution brought about low-voltage operation thanks to the high permittivity of the relaxor-ferroelectric terpolymer ($\kappa \approx 40$). Finally, the choice of an organic bilayer dielectric within this platform also benefitted from the wide applicability of polymer insulators as gate dielectrics in metal-oxide-semiconductor TFTs (see Chapter 2).

Oxide-only amplifiers reported to date all resort to vacuum-deposited gate dielectrics (PECVD SiN_x [6], PECVD SiO_x [133] and ALD AlO_x [122], [123], [132], [140]). The use of vacuum-processed dielectrics in these works can be understood as being the conventional choice of the metal-oxide community, in spite of the emergence of solution-processed inorganic dielectrics and the wide compatibility with polymer insulators (see Chapter 2). The high κ of most of the selected inorganic dielectrics (SiN_x and AlO_x) together with the excellent electronic properties of oxide semiconductors and their interfaces enabled low-voltage operation in the corresponding all-oxide amplifiers.

3.3.2 Evolution of Main Performance Parameters

The particular choice of materials, technologies and architectures has a profound impact on the performance parameters of flexible analogue amplifiers. Here we review the evolution and state of the art of these parameters, first focusing on gain and bandwidth, and then examining power supply voltage, common-mode rejection ratio, offset voltage and noise.

Amplification Gain

Gain values of all technology platforms evidence a sharp divide between unipolar and complementary technologies (see Figure 3.6a,

---

[14] Polymer semiconductors generally count a number of orthogonal solvents, and are thus compatible with a wider range of gate-dielectric inks (in a TG device stack). Small-molecule semiconductor films, instead, are easily dissolved or washed away by organic solvents, and most often are only compatible with gate-dielectric inks formulated in hyperfluorinated solvents.

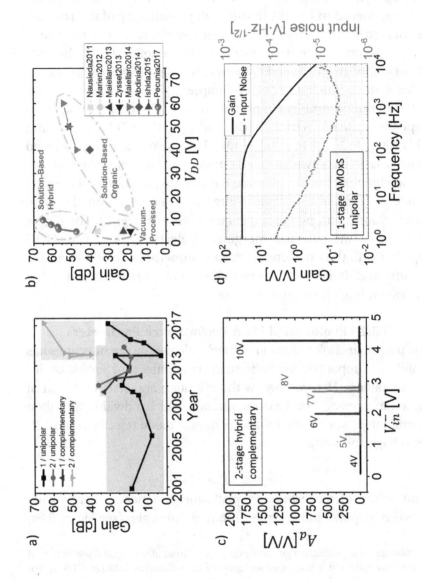

Caption for Figure 3.6 (cont.)

Figure 3.6 a) Gain of differential amplifiers (1 and 2 in the legend denote the number of stages) in organic and metal-oxide technologies over the years. Data points correspond to the journal articles listed in Table 3.1 and Table 3.2, and the associated gain values are the maximum reported in each of these publications. b) Gain versus power supply voltage in all two-stage organic and/or metal-oxide differential amplifiers in the literature (V. Pecunia, M. Nikolka, A. Sou, I. Nasrallah, A. Y. Amin, I. McCulloch, H. Sirringhaus, Adv. Mater. 2017, 1606938. Copyright Wiley-VCH Verlag GmbH & Co. KGaA. Reproduced with permission). The trend indicates the essential role of technology in achieving low-voltage operation. c) Differential gain versus input voltage of the two-stage complementary hybrid amplifiers in [8] at varying power supply voltages (V. Pecunia, M. Nikolka, A. Sou, I. Nasrallah, A. Y. Amin, I. McCulloch, H. Sirringhaus, Adv. Mater. 2017, 1606938. Copyright Wiley-VCH Verlag GmbH & Co. KGaA. Reproduced with permission). This implementation allows the lowest power supply voltage (5 V) while achieving the highest gain of all reports of two-stage flexible amplifiers to date. d) Frequency response (magnitude) and noise spectrum of the one-stage unipolar differential amplifier in [133], which features the highest gain of all one-stage unipolar flexible amplifiers reported to date.

Table 3.1 Implementations of organic and metal-oxide differential amplifiers in unipolar technologies. Organic amplifiers are followed by metal-oxide ones, each class being listed in chronological order. Lead authors of the articles are also indicated. (U): unipolar; Vacuum: vacuum-based; Solution: solution-based; N.A.: not available.

Articles	Technology Architecture	Substrate	Processing & Materials	V_{DD} (V)	No. of stages	Gain (V/V)	Power (µW)	f_{3dB} (Hz)
Kane 2000 [1][134]	OS (U) *zero-vgs depletion load*	PEN (75 µm)	Vacuum: sublimed pentacene, sputtered SiO_x	30	1	8.5	N.A.	N.A.
Gay 2006–7 [2][119]	OS (U) *diode-connected load*	silicon and glass	Solution/Vacuum: sublimed pentacene, spin-coated P4VP	40	1	2.5	N.A.	600
Marien 2010–1 [3][5][117]	OS (U) *BGE*	plastic foil	Solution: spin-coated pentacene, P4VP	15	1 1	2.5 5.6	N.A.	1.2 k 1.8 k
Nausieda 2011 [118]	OS (U) *zero-vgs load with dual V_T*	N.A.	Vacuum: sublimed pentacene, CVD parylene-C	5	2	63	$1.7 \cdot 10^{-3}$	8

Marien 2012 [4]	OS (U) *BGE*	plastic foil	Solution: spin-coated pentacene, P4VP	15	2	10	225	2 k
Chang 2014 [120]	OS (U) *diode-connected load with positive feedback*	polycarbonate	Solution: slot-die-coated TIPS-pentacene, screen-printed Dupont 5018	60	1 1	22.4 2.5	N.A.	4 10
Fukuda 2015 [121]	OS (U) *diode-connected load*	PEN (125 µm)	Solution/Vacuum: printed DTBDT-C6, CVD Parylene-C	30	1	4	15	N.A.
Tai 2012 [6]	AMOxS (U) *resistive load, backgate control*	Rigid	Vacuum: sputtered IGZO, PECVD SiN$_x$	10.5	1 1 1	2.8 3.4 9.8	N.A.	≈ 1 k ≈ 1 k ≈ 300
Zysset 2013 [122]	AMOxS (U) *diode-connected load*	PI (50 µm)	Vacuum: sputtered IGZO, ALD AlO$_x$	5	2	8.6	900	55 k

Table 3.1 (cont.)

Articles	Technology Architecture	Substrate	Processing & Materials	V_{DD} (V)	No. of stages	Gain (V/V)	Power (μW)	f_{3dB} (Hz)
Salvatore 2014 [123]	AMOxS (U) diode-connected load	Parylene (1 μm)	Vacuum: sputtered IGZO, ALD AlO$_x$	12	1	1.4	N.A.	$\approx$ 1 M
Ishida 2015 [140]	AMOxS (U) pseudo-CMOS load	PI (50 μm)	Vacuum: sputtered IGZO, ALD AlO$_x$	5	2	13.3	160	5.6 k
Shabanpour 2015 [132]	AMOxS (U) load with positive feedback	PI (50 μm)	Vacuum: sputtered IGZO, ALD AlO$_x$	6	1	8.9	6780	25 k
Garripoli 2017 [133]	AMOxS (U) diode-connected load plus positive feedback	PEN	Vacuum: sputtered IGZO, PECVD SiO$_x$	20	1	35	0.4	150

Table 3.2 Implementations of flexible differential amplifiers in complementary technologies (in chronological order). (C): complementary; Vacuum: vacuum-based; Solution: solution-based; †: compensated; N.A.: not available.

Articles	Technology	Substrate	Processing & Materials	V_{DD} (V)	No. of stages	Gain (V/V)	Power (μW)	f_{3dB} (Hz)
Vaidya 2010 [126]	OS (C)	PEN (75 μm)	Vacuum: sublimed pentacene, C_{60}, ALD AlO_x	12	1	44.7	0.6	N.A.
Guerin 2011 [127]	OS (C)	PEN (125 μm)	Solution: screen-printed acene-based diimide, PTAA, fluoropolymer dielectric	> 20	1	13.2 22,4	N.A. 40	1.02 k 0.6 k
Ishida 2012 [128]	OS (C) *floating-gate input pair*	PI	Vacuum: sublimed DNTT, NTCDI, CVD Parylene	20	1	4.9	N.A.	100
Maiellaro 2013 [7]	OS (C)	PEN (125 μm)	Solution: screen-printed Polyera ActivInk, TIPS-pentacene, and CYTOP®	50	2	355	300	< 0.1†

Table 3.2 (cont.)

Articles	Technology	Substrate	Processing & Materials	V_{DD} (V)	No. of stages	Gain (V/V)	Power (µW)	f_{3dB} (Hz)
Maiellaro 2014 [124]	OS (C)	PEN (125 µm)	Solution: screen-printed Polyera ActivInk, TIPS-pentacene, CYTOP®	40 50 60	2 2 2	224 354 501	N.A. N.A. 378	< 0.1†
Abdinia 2014 [125]	OS (C)	PEN (125 µm)	Solution: screen-printed acene-based diimide, PTAA, fluoropolymer dielectric	40	2	100	36	N.A.
Pecunia 2017 [8]	OS/AMOxS (C)	PI (3.8 µm)	Solution: spin-coated IDT-BT, IZO, high-κ polymer dielectric	5 8 10	2 2 2	233 1096 1798	11 107 135	N.A.

Table 3.1 and Table 3.2). Most unipolar technologies have gains below 10 V/V per stage [1]-[6], [117], [119]-[123], [132], [134], [140], and typically in the region of $2 - 5$ V/V. Many complementary platforms, instead, are capable of a gain per stage in the 20 − 40 V/V range [7], [8], [124], [126], [127]. Consistently with the models presented in Chapter 2, no correlation exists between gain and semiconductor field-effect mobility, the latter being the parameter primarily employed to evaluate flexible transistor technologies.

Unipolar technologies lead to gain figures below 10 V/V per stage for a large number of circuit architectures: with purely resistive loads [6], diode-connected loads [2], [119]-[123], zero-vgs depletion loads [1], [134], pseudo-CMOS loads [140], bootstrapped loads [3]-[5], [117] and loads with positive feedback [132]. The only unipolar works that notably surpass the 10 V/V threshold rely on a zero-vgs strategy (an approach generally burdened with high sensitivity to process parameter variations), thus attaining a gain per stage of 15 V/V [118]. Alternatively, they utilise some ingenious gain-enhancement architectures: Chang et al. and Garripoli et al. achieved gains of 22 V/V and 35 V/V per stage, respectively, by applying positive feedback schemes [120], [133]. In fact, the use of positive feedback and enhanced source-degenerated loads in dual-gate unipolar technology by Garripoli et al. has led to the highest gain of all one-stage unipolar flexible amplifiers reported to date (Figure 3.6a and Figure 3.6d).

The low gain figures produced by most unipolar amplifiers, both organic and metal-oxide, are a direct result of the difficulties related to the design of the load element in unipolar technologies. In this case, reliability considerations most often dictate that a diode-load is adopted, leading to gain values simply fixed by transistor sizing and at least an order of magnitude lower than the intrinsic transistor gain. While some more complex architectures afford higher gains, they are unfortunately burdened with increased transistor count and area consumption.

Complementary platforms have generally attained much higher gains (Figure 3.6a). This comes as no surprise, as the gain that can be reliably achieved by complementary technologies is not bound by geometry, and can approach the intrinsic gain of component transistors – nominally as large as $100 - 400\,\text{V}/\text{V}$ (see Chapter 2). The technological breakthrough in solution-processed complementary platforms has led to two-stage differential amplifiers with outstanding gain values: the all-organic work of Maiellaro et al. gave gain as high as $200 - 500\,\text{V}/\text{V}$ [7], [124], while the hybrid organic/metal-oxide implementation of Pecunia et al. reached $1800\,\text{V}/\text{V}$ (Figure 3.6c), the highest ever reported of all flexible amplifier implementations (Figure 3.6a) [8]. In terms of circuit topologies, Maiellaro et al. were the first to demonstrate cascoding in a printed organic amplifier [7], and showed that high gain figures per stage can indeed be achieved with this arrangement. Finally, in view of the nominal intrinsic gain figures of organic and metal-oxide technologies (see discussion in Chapter 2), differential complementary amplifiers could, in fact, potentially achieve even higher gains than those reported to date (i.e., in the range of 10^4 V/V in a two-stage architecture). However, note that: a) the nominal intrinsic gain figures in the literature were extracted from measurements on individual transistors; b) intrinsic gain is closely related to the non-idealities of TFT characteristics. Given the complexity of complementary processing, achieving the nominal intrinsic gain values in a complementary circuit constitutes a substantial challenge, which could however be addressed as the technologies mature.

Bandwidth

With few exceptions, the bandwidth f_{3dB} of organic amplifiers is typically in the range of $100\,\text{Hz} - 1\,\text{kHz}$ [2]–[5], [117], [119], [127], [128], [133]. These values refer to the amplifiers' intrinsic dynamic response, and are relevant to their use as open-loop gain blocks. When amplifiers are employed in feedback circuits, instead, it is usually necessary to introduce an ad hoc compensation network to ensure closed-loop stability. The work of Maiellaro et al. is the only case of solution-processed amplifiers where compensation was demonstrated [7], leading to a gain-bandwidth product of 75 Hz.

Metal-oxide amplifiers generally achieve higher bandwidths than organic ones, typically in the region $1 - 50$ kHz [6], [122], [132], [140]. A bandwidth as high as 1 MHz was achieved in [123], at a gain of only 1.4 V/V, however. The higher speed of metal-oxide amplifiers is reflective of the higher mobility of metal-oxide semiconductors, which translates into a larger transition frequency of the corresponding TFTs (Chapter 2).

The reported bandwidth values indicate that organic and metal-oxide amplifiers are sufficiently fast for relevant applications (i.e., smart-sensor systems, usually involving signal frequencies not higher than the kilohertz range). This is enabled by the mobility values of these semiconductor technologies, which are well suited for the resulting amplifiers to operate in the kilohertz range.

Power Supply Voltage

Power supply voltage is not an analogue amplifier's figure of merit per se, yet it is of fundamental importance when assessing the suitability of an amplifier for real-world applications. This is in view of the highly attractive potential applications involving portable and autonomous systems, in which the available power supply voltage amounts to a few volts. Many reports on organic and metal-oxide transistor technologies, however, involve bias voltages of several tens of volts (Figure 3.6b). Low-voltage operation has been frequently explored at the individual transistor level by the organic and metal-oxide communities, but remains largely unexplored at the circuit level. Low-voltage operation is in fact particularly challenging in amplifier circuits (as opposed, e.g., to digital logic gates), where all transistors are typically *on* and in saturation, thus all taking a finite share of the supply voltage.

The obvious trend that appears from organic amplifiers is that power supply voltages employed to date are most often in the range $15 - 50$ V, regardless of the specific transistor technology (Figure 3.6b). The only all-organic amplifiers that operate at reduced power supply values are those of Vaidya et al. (12 V) [126] and Nausieda et al. (5 V) [118]. It should come as no surprise that these works feature vacuum-sublimed small-molecule semiconductors and vacuum-deposited dielectrics, and thus rely on superior control of the active materials' electronic properties.

Most reports of metal-oxide amplifiers employ ≈ 5 V power supply voltages [6], [122], [132], [140]. It is noteworthy that all these implementations resort to sputter-coated semiconductors, which generally achieve electronic properties superior to the solution-processed counterparts. Additionally, these implementations use high-κ dielectrics to the end of increasing the gate-channel coupling, leading to the desired reduction of operating voltage.

A recent breakthrough in low-voltage operation was achieved through a solution-processed complementary platform featuring a polymer semiconductor, an innovative trap-healing approach, a high-κ polymer dielectric and precursor-based IZO [8]. Operation with a battery-compatible 5 V power supply was attained alongside a power dissipation of 11 μW and a gain of 233 V/V (Figure 3.6c). This same platform also achieved the highest gain figures reported to date for flexible amplifiers (> 60 dB) at a power supply voltage of only 8 V (Figure 3.6c). This indicates that flexible amplifiers based on organic and metal-oxide semiconductors are now approaching the power supply and power dissipation requirements of smart-sensor systems.

Common-Mode Rejection Ratio

Common-mode rejection is a key capability of differential amplifiers, making them more robust to noise and interference than single-input counterparts. Surprisingly, however, experimental evaluation of the common-mode rejection ratio (*CMRR*) of flexible amplifiers has not received much attention in the literature. The achieved *CMRR* figures range from 12 dB $-$ 20 dB for organic amplifiers (see Marien et al. [3], [117], and Nausieda et al. [118]), up to 20 dB $-$ 30 dB for the hybrid organic/metal-oxide complementary platform of Pecunia et al. [8] and to more than 40 dB for the unipolar metal-oxide amplifiers of Zysset et al. [122].

It is important to remind the reader that *CMRR* is heavily dependent on transistor variability and non-idealities. Therefore, the highest values were understandably achieved via a vacuum-processed unipolar technology (sputter-coated IGZO [122]), which allows greater device uniformity and better control of TFT non-idealities.

While greater emphasis on the experimental evaluation of common-mode rejection is desirable in the future, the *CMRR* levels reported thus far would already provide reasonable immunity from interference, hence are encouraging for applications in smart-sensor systems.

Offset Voltage

In principle, a zero signal at the output of a differential amplifier should correspond to a zero input signal. Non-idealities, however, may cause a non-zero input signal for a zero output, which defines the amplifier's input offset voltage. A large offset voltage is obviously undesired, because, for instance, it may compromise an amplifier's ability to discriminate the small signal from a sensor. Large input offset voltages have been generally reported for organic amplifiers, ranging from a few hundred millivolts [118] to several volts [1], [7], [124], [127], [128] (data on the offset voltage of metal-oxide amplifiers are generally lacking, however). The underlying cause is the mismatch between devices in the two branches of the differential amplifier, which is directly related to the substantial variability traditionally affecting flexible transistor technologies (see Chapter 2).

Different strategies have been proposed or pursued to obviate this problem. At a device level, modified transistor architectures were considered to compensate for threshold voltage variability: the addition of a floating gate into which charge would be injected [128]; double-gate structures requiring extra biasing of the additional gate terminal [6]. At the circuit-design level, Marien showed from simulation that diode-connected and bootstrapped loads give smaller offset voltages than zero-vgs loads [117]. Additionally, an auto-zeroing approach requiring a number of additional capacitors and switches has been proposed, which could in principle cancel the offset completely [143] yet could lead to reduced yield [144]. Finally, high-pass filtering has been successfully utilised to

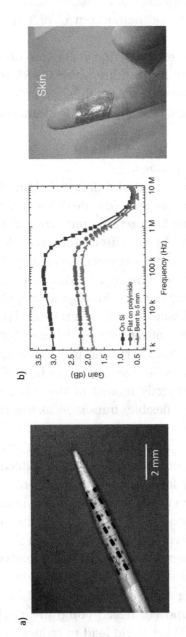

Figure 3.7 a) Hybrid analogue integrated sample on ≈ 3 μm thick PI foil bent over the surface of a toothpick (bending radius < 1 mm), realised as in [8]. b) Frequency response of the diode-connected-load unipolar IGZO amplifier in [123] fabricated on a 1 μm thick parylene film supported by a silicon carrier, after lamination on a 50 μm thick PI substrate, and upon bending to a radius of 5 mm (left); flexibility demonstration with the circuit placed on human skin (right). Adapted with permission from Macmillan Publishers Ltd: Nature Communications ([123]). Copyright (2014).

prevent the amplification of the offsets through the chain of amplifiers in a multistage configuration [117].

As far as the available data allow us to conclude, offset voltage remains a problem in flexible analogue amplifiers. The authors believe that the answer is likely to require further efforts at the processing level, aiming at the development of techniques that could make organic and metal-oxide semiconductors more resilient to variability.

Noise
High TFT noise makes it difficult to achieve amplifiers with low input-equivalent noise, as needed in smart-sensor front-ends. For example, the amplifier in [145] achieves an integrated input-referred noise in band of 2.3 µV root mean square (RMS) (100 Hz bandwidth), at the expense of large power dissipation (11 mW), high supply voltage (55 V) and large area occupation ($> 2\,cm^2$). An input-referred noise of 146 µV (RMS) was measured in the IGZO amplifiers in [133], while simulations showed that this value could be reduced to below 30 µV (RMS) in a noise-optimised design. Generally, flicker noise is dominant in flexible circuits due to their operation at low frequencies, hence the use of flicker noise reduction techniques such as chopping [146] is likely to be necessary for sensor applications.

3.3.3 Flexibility
Besides circuit performance figures, a key aspect of flexible analogue amplifiers is, by definition, their flexibility. Amplifiers in the literature address this demand by utilising flexible plastic substrates (e.g., PEN and PI). Foil thickness is most typically at 50 − 125 µm [1], [7], [121], [122], [124]-[127], [132], [134], [140] − which affords flexibility down to a bending radius of a few millimetres − and can go down to 1 − 3 µm thickness to attain submillimetre bending radius and increased lightness (in the realm of imperceptible electronics, see Figure 3.7a) [8], [123]. The specifics of the amplifier behaviour under bending have not captured much attention in the literature, and need to be investigated more

carefully. The only experimental evaluation reported to date of an amplifier under bending is provided in [122] and [123], and relates to AMOxS unipolar diode-connected-load amplifiers. At a bending radius of 5 mm (0.5 per cent strain parallel to the TFT channel), the change in the amplifier frequency response (hence, gain and unity-gain frequency) reported in both works is extremely small (e.g., see Figure 3.7b). This behaviour can be traced to the diode-connected-load architecture, in which gain is primarily fixed by TFT geometry.

In general, the advancement of organic and metal-oxide amplifiers would greatly benefit from an extensive evaluation of their behaviour under bending in the manifold architectures and technologies pursued to date. In fact, the response of an amplifier depends closely on the detailed shape of its component transistors' characteristics, much more so than digital circuits. Additionally, organic and metal-oxide TFTs do undergo slight changes in device characteristics under strain (e.g., mobility can shift by a few percentage points in metal-oxide and organic polymer TFTs, and up to a few tens of percentage points in small-molecule organic TFTs [42]). The experimental assessment of amplifier behaviour under strain would thus help pinpoint circuit architectures and technologies that are more promising for flexible amplifier applications.

3.4 Concluding Remarks

This chapter has discussed in detail amplifiers based on organic and metal-oxide semiconductors, which constitute a key circuit block for flexible analogue electronics. It has been shown that tremendous progress has been achieved over the years, especially in terms of gain and low-voltage operation, and the current performance levels are now promising for sensing applications. The discussion now shifts to another key element needed for sensorisation, namely ADCs, which are essential for the digitisation of the signals from the amplifier circuits. ADCs are discussed in Chapter 4.

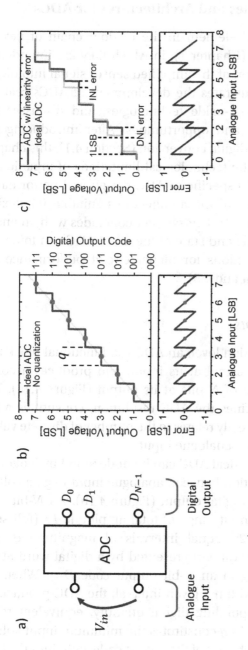

Figure 4.1 a) Block diagram of an N-bit ADC. b) Ideal transfer function and corresponding quantisation error ($N = 3$ bits). c) Nonlinearity and associated error.

4   Circuit Techniques and Architectures for ADCs

ADCs are an essential element in the readout chain of flexible smart-sensor systems (Chapter 1), in which they are intended to carry out the conversion of the amplified sensor signal into digital form. This chapter illustrates the development of ADCs realised with organic and metal-oxide technologies, and discusses the associated challenges and opportunities. After introducing the basics of analogue-to-digital conversion (Section 4.1), this chapter gives an overview of the technological limitations of organic and metal-oxide electronics specific to ADC design and performance (Section 4.2). A survey of the architectures suitable for flexible ADCs follows (Section 4.3). This chapter concludes with an analysis of the developments and state of the art of ADCs on foil, from which promising directions for higher circuit performance are identified (Sections 4.4 and 4.5).

4.1   General Concepts

At the highest abstraction level, an ADC is a functional block that receives an analogue signal at its input, and produces a corresponding digital word of N bits at its output (Figure 4.1a). The number N, which defines the ADC resolution, constitutes a key figure of merit, as it directly relates to the number of discrete values utilised to represent the analogue input.

The operation of an ideal ADC can be understood by looking at its static transfer function between analogue input (e.g., a voltage in this example) and digital output (Figure 4.1b). An N-bit ADC divides its analogue input range of total amplitude FS (full-scale voltage range) into 2^N equal intervals of magnitude q (i.e., $q = FS/2^N$). Each interval is represented by a digital word at the ADC output, resulting in an N-bit output code D_{out}. When the analogue input falls within a given interval, the ADC produces at its output the corresponding digital quantity, equivalent to an output voltage $V_{D,out}$. As q constitutes the minimum input voltage change resulting in a change of the output code, it defines the least significant bit (*LSB*) voltage. This conversion process, known as *quantisation*, inherently introduces an error $V_{in} - V_{D,out}$ between the analogue input and its digital representation (quantisation error). By inspection of Figure 4.1b, it is apparent that the maximum quantisation error in an ideal ADC amounts to $q/2$ (i.e., *LSB*/2).

The transfer function of a real ADC generally deviates from the ideal situation of Figure 4.1b. The resulting nonlinearity is specified as differential and integral nonlinearity (*DNL* and *INL*, respectively) expressed in unit of *LSB*. Differential nonlinearity represents the difference between the width of a given input step and its ideal value of $1\,LSB$ (Figure 4.1c). Integral nonlinearity, instead, characterises the discrepancy between the location of a given input step and its ideal value (Figure 4.1c). As a result of nonlinearity, the actual quantisation error exceeds the ideal $LSB/2$ value.

Two key aspects of ADC performance, i.e. accuracy and speed, are illustrated in the rest of this section.

4.1.1 Accuracy

An obvious requirement for ADCs is that they provide an accurate representation of the input signal. While carrying out the digitisation of the input signal, an ADC adds noise to it. This can be modelled as an equivalent noise source at its input, and should be kept as small as possible. In fact, in a smart-sensor system (Figure 1.2), this noise sets a limit for the accuracy of the full readout electronics chain, which cannot be better than the one offered by the ADC.

Similarly to amplifiers, ADC noise performance can be expressed in terms of signal-to-noise ratio, defined as the ratio (in dB) between the signal power P_{signal} and noise power P_{noise}:

$$SNR_{dB} := 10 \log_{10} \frac{P_{signal}}{P_{noise}}$$

Noise contributions in ADCs arise from several sources, and most importantly:

a) inherent quantisation error, leading to the so-called quantisation noise, which can be shown to have an RMS value of $q/\sqrt{12}$;[15]
b) nonlinearity of the transfer function, further increasing the RMS quantisation noise;
c) physical noise of devices (e.g., TFT noise).

Besides, interference and instabilities may limit the input signal level that can be processed by the ADC to below the maximum full-scale ($|V_{in,max}| < FS$), deteriorating, thus, the SNR.

As the input signal is generally time-variant, signal-dependent distortions also cause deviation from the ideal ADC response. Key manifestations are:

a) harmonic distortion due to nonlinearities (i.e., appearance of harmonics of the input signal in the output spectrum);
b) signal-dependent behaviour of quantisation noise;
c) distortion dependent on input signal frequency.

The effect of distortions in an ADC are included in its signal-to-noise-and-distortion ratio ($SNDR$):

$$SNDR_{dB} := 10 \log_{10} \frac{P_{signal}}{P_{noise} + P_{distortion}}$$

where $P_{distortion}$ refers to all output spectral components not due to noise and nonlinearly related to the input signal. An equivalent figure of merit is the effective number of bits ($ENOB$), a parameter defined as:

$$ENOB := \frac{SNDR_{dB} - 1.72}{6.02}$$

While the physical number of bits N of an ADC determines its upper resolution limit, actual noise and distortion lead to a reduced resolution equal to $ENOB$. It is thus apparent that, to

---

[15] For a full-scale sinusoidal input signal, this leads to $SNR_{dB} = 6.02\,N + 1.76\,(\text{dB})$.

improve the accuracy of an ADC, a larger number of bits is needed and/or its distortion and noise sources must be minimised.

4.1.2 Speed

An ADC is required to digitise the input signal at a sufficiently fast rate to allow its reconstruction at the output. This typically sets an upper limit on the bandwidth f_{in} of the analogue input signal that can be processed reliably: indeed it must be true that $f_s \geq 2 f_{in}$, where f_s is the rate at which the input signal is digitised (Nyquist sampling theorem). In so-called Nyquist-rate ADCs $f_s = 2 f_{in}$, while an ADC is said to be oversampling the input signal if $f_s > 2 f_{in}$, and, correspondingly, an oversampling ratio OSR is defined, $OSR := f_s / (2 f_{in})$. The timing of the digitisation process can be set by an external periodic clock signal of frequency f_{clk} (synchronous sampling), and digitisation may take place over one or more clock periods.

4.2 Technological Challenges for Organic and Metal-Oxide ADCs

The realisation of ADCs using organic and metal-oxide technologies presents significant challenges, which primarily relate to a) the common lack of a complementary integration platform; b) variability, yield losses and environmental instability; c) speed; and d) supply voltage and power dissipation. These challenges need to be considered in relation to the mixed-signal nature of ADCs, which, in fact, comprise both analogue and digital circuit blocks (Section 4.3). Here we give an overview of each of these challenges and assess their criticality level.

ADC integration using a unipolar technology is considerably more challenging than on a complementary one. Complementary technologies inherently lead to amplifiers with larger amplification gain and potentially higher gain-bandwidth product, which, for instance, could be used to improve ADC accuracy (Section 4.3). Complementary platforms also facilitate the use of switched-based analogue techniques (e.g., chopping, auto-zeroing, etc.), typically employed for noise reduction and offset cancellation. Unipolar circuits, instead, generally demonstrate lower gain and gain-bandwidth product. Additionally, achieving large gain via multistage unipolar amplifiers may come at the cost of using level shifters or bulky coupling capacitors [11], [117]. Organic and metal-oxide technologies were traditionally unable to deliver robust complementary platforms. Therefore, researchers pursued alternative routes to mitigate the limits of unipolar technologies, and appreciable performance improvement was achieved, for instance, with a dual-gate unipolar approach. Recent years have witnessed the emergence of robust complementary platforms [7], [8], [139], however, which bring along new opportunities for improved ADC design and performance.

The availability of a complementary platform (or lack thereof) has also been an important aspect in the design of the flexible ADCs' digital sub-blocks. Complementary digital circuits are widely recognised for their superior noise margins, gain and robustness, on top of their negligible static power dissipation. In contrast, digital circuits in unipolar technologies are inherently affected by appreciable power dissipation. Additionally, in their basic architectures (e.g., diode-connected or zero-vgs loads), they lead to substantially lower noise margins, and manifest higher sensitivity to process parameter variation. Consequently, ADC designers working with unipolar platforms had to develop a range of circuit techniques to achieve adequate noise margins and robustness to TFT variability [103], [130].

Yield loss, variability and environmental instability (see Chapter 2) are significant sources of concern in ADC design, as they directly affect functionality and performance. For instance, an important contribution to ADC nonlinearity arises from the mismatch between nominally equal sub-blocks, which, in turn, results from the mismatch between component devices (e.g., TFTs, resistors and capacitors). When larger devices are used to reduce mismatch (with a standard deviation inversely proportional to the square root of device area), the circuit might become more vulnerable to hard faults. The variability issues of organic and metal-oxide TFTs have thus forced designers to develop techniques and architectures capable of enhancing soft yield. Although the use of a complementary technology typically improves soft yield, the complexity of integrating both p- and n-channel TFTs could potentially deteriorate hard yield.

In addition to device mismatch, the shift in TFT parameters from one batch to another and/or under environmental conditions implies that ADCs on foil are required to be performant over large parameter ranges. In general, to boost robustness, designers typically aim to keep the circuit design as simple as possible, and choose for circuit topologies that are robust to shifts in TFT characteristics. Availability of data on variability, hard yield and reliability can help designers in making the right choices for circuit and system architecture.

The limited carrier mobility of organic and metal-oxide semiconductors is bound to result in ADCs with limited speed, especially in comparison with conventional IC-based counterparts. However, this does not constitute a major challenge for the adoption of organic and metal-oxide ADCs in smart-sensor systems. Indeed, relevant applications often involve quasi-static measurements (of parameters such as temperature, pressure, etc.), whose speed requirements are within reach of organic and metal-oxide technologies.

A final element of particular importance for organic and metal-oxide ADCs is their power supply voltage and power dissipation. These are essential issues in view of the ADCs' intended application to battery- or wirelessly powered smart-sensor systems. The use of complementary technology can improve ADCs' performance in terms of required power consumption. Additionally, ADCs should operate at a sufficiently low power supply voltage (i.e., 5 V), yet the reported operation of individual organic and metal-oxide TFTs often involves a few tens of volts.

Despite all these challenges, researchers have demonstrated a few examples of ADCs on foil integrated using different technology platforms. This has been enabled, on one hand, by a technology-

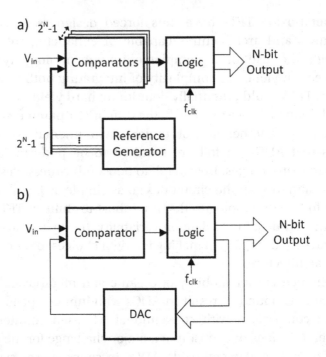

Figure 4.2 ADC block diagrams. a) Flash ADC. b) Single-comparator DAC-based ADCs (e.g., SAR ADC, counting ADC, etc.).

aware circuit design, which carefully takes into account the characteristics of each technology. On the other hand, material and process developments have themselves helped to relieve or overcome some of these challenges.

4.3   Architectures for ADCs on Foil

In this section we present different architectures that are of relevance to ADCs on foil. Besides illustrating the operating principles of each architecture, a critical assessment is provided on their suitability for integration on foil, in the light of the technological challenges described in Section 4.2.

4.3.1 Basic Architectures

Based on the static characteristics in Figure 4.1b, an ADC divides the full-scale input range into equal sub-ranges, and determines to which sub-range the input belongs. To perform this function, the ADC generates analogue references and compares the input signal with the reference levels, using comparators. Based on the comparison result, the ADC produces an output code, using a combination of logic gates.

In a flash ADC, Figure 4.2a, all the reference voltages are generated and compared with the analogue input simultaneously. Several comparators ($2^N - 1$ in an N-bit ADC) work in parallel, and the comparator outputs can be directly combined to generate the output code. It is clear that this approach sets stringent requirements on comparator variability. In particular, offset variability directly translates into nonlinearity of the ADC transfer function. Additionally, the large number of comparators required implies a high area consumption and, consequently, the demand for very good hard yield. For these very reasons, this approach has been never pursued for organic and metal-oxide technologies on foil.

In ADC architectures based on a single comparator, Figure 4.2b, the comparison between input and reference levels is performed in multiple successive steps, rather than in a single step (as in the flash ADC), exploiting in a suitable way a digital-to-analogue converter (DAC). Successive approximation (SAR) and digital slope (counting) ADCs belong to this group. In these single-comparator DAC-based ADCs, the DAC creates an analogue approximation of the input voltage, based on the digital output, while the logic changes the digital output code according to a given algorithm. The algorithm finds the best possible approximation of the input, based on the information from the comparator. The multi-step conversion makes it possible to reuse the same hardware for all the conversion steps, though at the expense of a lower throughput. As a result, the requirements on matching and variability are relaxed, making these architectures attractive for the integration of organic ADCs on foil [147], [148]. For example, SAR and counting

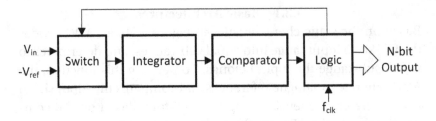

Figure 4.3 Block diagram of a dual-slope integrating ADC.

ADCs are insensitive to comparator offset, since this remains the same in all steps and only results in an offset error in the ADC transfer function. In these architectures, the accuracy of the ADC is limited mainly by nonlinearity in the reference generation, i.e. the nonlinearity of the DAC.

A DAC typically relies on matched devices[16] (e.g., resistors, TFTs, capacitors) to divide a reference quantity (e.g., voltage or current) into smaller levels, used as references by the ADC. Device mismatch would thus lead to unequally spaced reference levels, directly translating into nonlinearity in the transfer function of the DAC and, consequently, the ADC. TFT matching in conventional photolitographic processes on foil is typically limited to about 10 per cent [149]. Passive elements (e.g., resistors and capacitors) are expected to have better matching, since their electric characteristics depend on fewer process steps and material properties. Indeed, capacitors manufactured via photolithography (area of $0.04\,\text{mm}^2$) and screen-printed resistors (area of $54\,\text{mm}^2$) demonstrated matching of 1.7 per cent [107] and 1 per cent [12], respectively. This level of matching is enough to provide an effective resolution (*ENOB*) of $6-7$ bit for the DAC and ADC, at the cost, however, of large area occupation [12], [107].

The lack of a linear reference generation can be addressed by an architecture called dual slope integrating ADC. The key idea is to use time-domain references, i.e. multiples of a clock period T,

---

[16] *Device matching* here refers to devices that are nominally identical and placed close by, so that their parameters are as similar as possible.

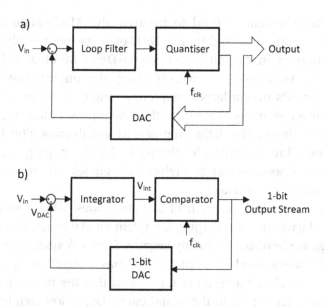

Figure 4.4 Synchronous delta-sigma modulator. a) General block diagram. b) Block diagram of first-order 1-bit implementation.

rather than voltage-domain ones. The basic block diagram of a voltage-input dual-slope integrating ADC is shown in Figure 4.3. First, for a constant number of clock cycles (N_1), the sampled input is integrated and the output of the integrator (V_{int}) ramps up. Then, using a constant negative voltage ($-V_{ref}$) as input, the integrator output ramps down with a constant slope, and the number of clock cycles required for V_{int} to reach zero (N_2) is counted, using a comparator and logic (counter). The output code representing the input voltage depends on N_1 and N_2 and is independent of T, provided that T is stable during the whole conversion time. Assuming that the integration process is sufficiently linear, the ADC accuracy here is primarily determined by the length of the integration period. With a full-scale input, a longer ramp-down period signifies a larger number of bits (provided that the limits of comparator sensitivity and integrator output swing are not reached).

Another classical method to improve the ADC resolution far beyond the one provided by the DAC exploits oversampling and noise shaping in the so-called Sigma-Delta Modulator (SDM) approach. As a result of oversampling, the quantisation noise power spreads over a larger frequency range, i.e. while the total noise power remains the same, the noise power density (noise floor) goes down. The SDM then acts as a high-pass filter for the noise and a low-pass filter for the signal (noise shaping), pushing most of the noise power to higher frequencies, where it can be filtered out at a later stage.

The general block diagram of a synchronous SDM is shown in Figure 4.4a, where noise shaping is implemented through a loop filter and negative feedback. As in any negative feedback system, the noise added by the second stage (quantiser) has a significantly smaller effect on the closed loop output, provided that the first stage (loop filter) has a large gain in the signal band (i.e., at low frequencies). Since accuracy is achieved by shaping the noise and filtering it out later, the requirements on the DAC are relaxed. In other words, the ADC resolution is no longer limited by the DAC, thus allowing DACs with a resolution as low as 1 bit to be employed. A 1-bit DAC is inherently linear and, thus, a suitable option for ADCs on foil.

A first-order implementation of the SDM concept is achieved by employing as sub-blocks an integrator (loop filter), a comparator (quantiser) and a 1-bit DAC, as depicted in Figure 4.4b. The difference between the input signal and the DAC output $(V_{in} - V_{DAC})$ is fed to an integrator, causing it to ramp up or down, with a slope and direction depending on $(V_{in} - V_{DAC})$. When V_{int} reaches the comparator's threshold voltage, the comparator flips and, at the next clock, changes the DAC's state. As a result, the integrator will start to ramp in the opposite direction. The clocked output of the comparator is a high-frequency pulse wave in which the ratio of the number of 'ones' to 'zeros' is proportional to the input signal, and the average of the pulse train is equivalent to the input value. In an SDM-based ADC, the e.g. 1-bit stream output of the modulator is then post-processed

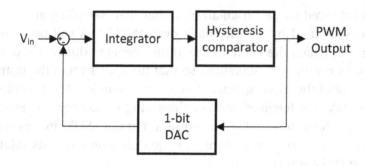

Figure 4.5 Block diagram of a first-order ASDM.

through filtering and down sampling, to generate the multi-bit digital representation of the analogue input.

To increase accuracy, the SDM takes advantage of high speed sampling together with high gain and, thus, needs amplifiers with large gain-bandwidth product to implement the loop filter. To give an example, if the input signal is sampled before the sigma-delta feedback loop (the so-called discrete-time delta-sigma modulator), the GBW of the amplifiers in the loop filter should be at least 5 times larger than the sampling frequency f_{clk}. As explained in Section 4.2, the gain-bandwidth product of amplifiers on foil is generally limited. Besides, the speed of components like the comparator is limited when implemented with AMOxS or OS TFTs. Consequently, SDM-based ADCs on foil cannot adopt a large oversampling rate, hence the SDM approach is not as effective as in mainstream technologies and the *ENOB* achieved by state-of-the-art sigma-delta ADCs is limited.

4.3.2 Alternative Data-Conversion Architectures

In Section 4.3.1, we saw that SDM-based ADCs do not rely on high-resolution DACs to achieve high accuracy, yet they require amplifiers with high gain-bandwidth product. This requirement is relaxed to some extent when a less-known SDM variant, the asynchronous sigma-delta modulator (ASDM), is adopted [9], [150].

The structure of an ASDM is similar to a synchronous SDM, but with fundamental differences [150]: the effect of oversampling is now achieved via self-oscillation, rather than sampling at the frequency defined by an external clock. ASDM is a self-oscillating circuit in which the value of the input alters the duty cycle of the output's rail-to-rail oscillation, so that the spectrum of the output reproduces the input spectrum at low frequencies. As in synchronous SDM, the feedback loop enables a high-accuracy conversion. Besides, the absence of synchronisation in the ASDM makes sampling unnecessary, and eliminates quantisation (with its related noise) in the ASDM loop.

As shown in Figure 4.5, ASDM can be implemented using an integrator (as a loop filter), a hysteresis comparator (as quantiser) and a 1-bit DAC. The output oscillation frequency is set by the moment the output of the integrator crosses the threshold of the comparator. The maximum output carrier frequency is known as the *limit cycle frequency*, f_c, which is an important parameter to determine the conversion quality [150].

In ASDMs, the input signal amplitude is converted to the time domain continuously and without loss of information, i.e. with no quantisation noise. This allows for a simpler and more relaxed circuit implementation [150]. Indeed, a low amplifier gain is sufficient to obtain good *SNR* [150], while the required amplifier bandwidth is in the range of f_c (compared to several times f_{clk} to allow for the circuit to settle with required accuracy in the synchronous discrete-time SDM). The ratio between the limit cycle f_c and the input frequency f_{in} (i.e., *OSR*) can be limited while keeping the linearity errors low, even for input signals close to maximum. Indeed, in ASDM, the spurious-free dynamic range (SFDR) can be larger than 70 dB for an *OSR* = 10 at a modulation depth of 80 per cent [150]. These characteristics are very promising for the implementation of ASDMs on foil, as the sensor signals are normally very slow and the gain-bandwidth product of flexible amplifiers is generally limited. Finally, as the ASDM is self-oscillating, it provides some form of self-adaptation to the characteristics of the implemented circuit. This last point can be very interesting in circuits implemented with TFTs on foil, which are affected by very large variability.

Table 4.1 State of-the-art ADCs realised with organic or metal-oxide semiconductors. (U): unipolar; (C): complementary; Vacuum: vacuum-based; Solution: solution-based; N.A.: not available.

Articles	Xiong 2010 [147]	Marien 2010–1 [5][117]	Raiteri 2013 [151]	Abdinia 2013 [148]	Garripoli 2017 [9]
Technology	OS (C)	OS (U)	OS (U)	OS (C)	AMOxS (U)
Semiconductor and Gate Dielectric Processing	Vacuum: sublimed DNTT, $F_{16}CuPc$, evaporated AlO_x	dual-gate Solution: spin-coated pentacene, P4VP	dual-gate Solution: spin-coated pentacene, P4VP	Solution: screen-printed Polyera ActivInk, TIPS-pentacene, CYTOP®	dual-gate Vacuum: sputtered IGZO, PECVD SiN_x
Substrate	glass	plastic foil	PEN (25 µm) on rigid carrier	PEN (125 µm)	PEN
L [µm]	20	5	5	20	15 (single-gate) 30 (dual-gate)
V_T [V]	−0.5 (p-type) 0.5 (n-type)	N.A.	1.2	−20 (p-type) 20 (n-type)	0.2
μ [cm² V⁻¹ s⁻¹]	0.5 (p-type) 0.02 (n-type)	1	0.01	1.5 (p-type) 0.5 (n-type)	14

Table 4.1 (cont.)

Articles	Xiong 2010 [147]		Marien 2010-1 [5][117]	Raiteri 2013 [151]	Abdinia 2013 [148]	Garripoli 2017 [9]	
Architecture	SAR*		SDM	VCO-based	Counting-based	ASDM	
Output type	Digital word		Bit stream	Digital word	Digital word	PWM	
Complexity (no. of TFTs)	53		70	~550****	148	~30	
Resolution (bit)	6	5 50	3.8**	6	4	8**	6.4**
BW (Hz)	N.A.		15.6	~0.1	2	10	300
SNR (dB)	N.A.		26.5	48	25.7	55	43
SNDR (dB)	N.A.		24.5	N.A.	19.6	50	40
INL (LSB)	0.6	1.5	N.A.	1	0.42***	N.A.	
V_{DD} (V)	3		15	20	40	20	
Power (µW)	3.6		1500	48	540	2000	
Area (mm²)	616		260	19.4	2180	27.9	

* Off-chip logic.   ** Evaluated from the *SNDR*.   *** From *INL* of DAC.   **** With 10-bit ripple counter.

It is important to note that the output of an ASDM is a pulse with a duty cycle (pulse width) proportional to the input amplitude, i.e. it is a pulse width modulated (PWM) signal. Even though a PWM pulse is not a digital word, it only has two amplitude levels and thus is robust enough to be easily and accurately transferred to a reader [9]. Therefore, in smart-sensor systems on foil, the complex decoding schemes required for conversion to a multi-bit digital representation can be performed at the reader side or in a silicon chip. In other words, using ASDM, the actual analogue-to-digital conversion can be shifted to the silicon circuitry, where high-performance transistors are available and high accuracy can be easily reached. The generation of a PWM representation of an analogue input, and in general voltage-time conversion, requires the power supply to be constant during the generation of the data stream. This could especially be challenging if the power is obtained from electromagnetic waves received on foil, like in RFIDs.

4.4 Developments and State of the Art

ADCs manufactured in organic and/or metal-oxide technologies have been reported only in a limited number of papers [9], [117], [147], [148], [151]. State-of-the-art ADCs on foil have been implemented in different architectures and on various process platforms. Due to the variety of processes employed, a direct comparison between different implementations is not always meaningful. The performance of ADCs is greatly affected by technology features (e.g., mobility and threshold voltage), and the range of architecture choices available is dependent on the lack (or availability) of dual-gate or complementary TFTs.

Table 4.1 gives an overview of key technology features, architectures and circuit performance metrics of organic and metal-oxide ADCs. In addition to semiconductor, dielectric and substrate materials and processing, channel length (L), threshold voltage (V_T) and mobility (μ) are included as main technology characteristics. Reported circuit specifications include resolution (physically

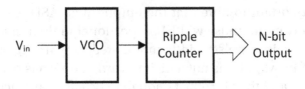

Figure 4.6 VCO-based ADC [151].

implemented number of bits, when applicable, otherwise *ENOB*), bandwidth (*BW*), signal-to-noise ratio (*SNR*), signal to noise and distortion ratio (*SNDR*), integral nonlinearity (*INL*), supply voltage (V_{DD}), power consumption and area.

One of the first demonstrations of an ADC made with OSs was produced by Xiong et al., in a complementary vacuum-based technology on glass, and adopted a SAR architecture [147]. The use of complementary transistors made it possible to implement the comparator and DAC based on simple CMOS switched-capacitor architectures, difficult to realise in unipolar technologies. The DAC relies on the matching of capacitors. The SAR logic in this case is off-chip and calibration techniques are used, leading to a linearity sufficient for close to 6-bit resolution. The SAR ADC consumes very low power, typical of the SAR architecture and facilitated by the use of a complementary technology.

Dual-gate organic TFT technologies were used by Marien et al. [117] and Raiteri et al. [151] to implement an SDM-based and an integrating-based ADC, respectively. The SDM uses a three-stage amplifier as integrator, incorporating different analogue circuit techniques to improve gain and speed, such as bootstrapped gain enhancement and backgate steering (see implementations in Section 3.2.3). The comparator also attains a high gain through a multistage configuration. In all these building blocks, the use of the backgate is of paramount importance to building block performance, resulting ultimately in an *SNDR* equivalent to a 3.8-bit resolution.

Raiteri et al. proposed a simple ADC architecture based on a voltage-controlled oscillator (VCO), shown in Figure 4.6 [151]. A linear VCO converts the analogue signal amplitude to frequency, which is then converted to a digital code by a counter. The fundamental operation is somehow similar to that of a dual-slope integrating ADC [11]: the conversion to time domain is done by VCO (T_{VCO}), and the integration during a fixed period (T_{INT}) is performed in digital domain by a ripple counter. The ADC output is proportional to T_{INT}/T_{VCO}, and the overall resolution depends on the integration time, the range of the VCO frequency ($f_{max} - f_{min}$) and its linearity. Two integration times are used for reference generation and one for signal conversion. Similar to the circuits in the aforementioned SDM-based ADC [117], the use of the second gate is fundamental for achieving the required linearity in the transconductor-based VCO. The period to jitter *SNR* of the VCO is 48 dB, and an overall ADC linearity of around 5 bit can be achieved for quasi-static input signals (T_{INT} of 6.2 s), without use of any calibration techniques.

When it comes to yield and area consumption, the VCO-based ADC stands out, due to its simplicity. For a long-enough conversion time, the design effort mainly regards the linearity of the voltage-time conversion. However, the ADC requires a stable time reference (T_{INT}) that is constant over the long measurement intervals, which can be sent through a radio link in an RFID platform. In addition, the dependence of accuracy on integration time implies that, for acceptable performance, the conversion speed is extremely limited, making the circuit suitable only for quasi-static applications.

Abdinia et al. used a printed complementary technology to implement a counting-based ADC on foil (the general architecture is shown in Figure 4.2b) [148]. A 4-bit R2R DAC based on printed resistors is accompanied by a switched-capacitor mismatch-free comparator, and a transmission-gate based counter, made possible by the availability of both p- and n-channel TFTs. The first measured *SNDR* is equivalent to 3-bit resolution for quasi-static signals, and it is limited by the DAC linearity. Further measurements on DAC samples manufactured later, after improvements in the technology, show that a linearity as high as 7 bits is attainable with this structure, thanks to the good matching between printed resistors [12]. The large area consumption is due to both the large resistors used and the large circuit footprints typical of the printed complementary technology (e.g., 7 mm^2 for an inverter). Although they may suffer from lower yield and larger area consumption and parasitics, printed circuits remain absolutely relevant because of the advantages they offer in terms of low cost and high throughput of the production process. The large area of the circuit is an indication of the capability of the printing-based technology to cover large area surfaces with functional electronics.

The use of ASDM for data conversion on foil was explored by Garripoli et al. [9]. In this ADC, the integrator amplifier employs positive feedback to achieve high gain (see implementation in Section 3.2.3). The comparator includes a preamplifier and a latch, both of which also use positive feedback to improve gain, exploiting dual-gate devices. An 8-bit *SNDR* is achieved in 10 Hz bandwidth, thanks to the absence of quantisation noise in the ASDM architecture. Besides high resolution, the ASDM ADC can reach a bandwidth of 300 Hz (at lower resolution), while all the other ADCs on foil function at a bandwidth of a few hertz (even lower in case of the VCO-based architecture). This is mostly due to the mobility of IGZO TFTs, which is at least an order of magnitude larger than that in the organic TFTs used in other circuits. Indeed, at a similar current consumption level and with smaller minimum feature size, the SDM ADC in organic technology described earlier [5], [117] reaches a bandwidth of 15 Hz.

4.5 Concluding Remarks

Despite the advancements in the design of circuits and systems on foil, the number of examples of ADCs on foil in the state of the art is relatively limited. The mixed-signal nature of ADCs has made it difficult to achieve the required specifications, due to the limitations that flexible TFTs have imposed on circuit design thus far. The generation of accurate reference voltages required in an ADC, which typically relies on device matching, seems to be a particularly limiting factor.

The specific characteristics of the technology in use have a key role in making decisions about the ADC architecture and other design aspects. It is noteworthy that all the state-of-the-art ADCs implemented on foil use either a complementary or a dual-gate technology. In the absence of complementary TFTs, the presence of a second gate seems indispensable for analogue circuit techniques to achieve the gain and linearity required for practical applications. Indeed, Table 4.1 shows that the ADCs in unipolar technologies adopt architectures in which linearity is highly dependent on the characteristics of organic TFTs, controlled by the backgate. ADCs using complementary TFTs tend to rely on the matching of passive elements and switch-based solutions to cancel the offset in the analogue sub-blocks.

The availability of a complementary technology can facilitate the implementation of simple and effective offset cancellation techniques, large gain-bandwidth amplifiers and high-yield digital circuitry. Therefore, it provides a larger degree of freedom in the choice of architectures for ADCs on foil. In addition, complementary ADCs have a potentially lower power supply voltage and power consumption, important for sensor-on-foil applications in which the circuit is expected to be powered through a battery or an RFID radio link.

An effective approach for high-accuracy conversion is to generate a PWM representation of the input signal rather than a digital word using an ASDM. This avoids the need for accurate voltage references, and relaxes the requirements for the analogue building blocks. In this case, analogue to digital conversion can be shifted to a back-end silicon chip. This technique is suitable for sensor systems that, in any case, require further data processing to be performed in a silicon chip in hybrid connection to the flexible electronics, or in a base station.

5 Conclusions

This Element has discussed the recent developments and state of the art of organic and metal-oxide analogue electronics.

Focusing on two prototypical circuit blocks, namely amplifiers and ADCs, this Element has illustrated that this area has known tremendous performance improvement in recent years, with important milestones being achieved in respect to a number of metrics. On one hand, the gain provided by organic and hybrid differential amplifiers has reached values unimaginable only a few years back – above 40 dB and even greater than 60 dB – thanks to complementary integration. Concurrently, operating voltages of solution-processed differential amplifiers have been reduced down to battery-compatible values of 5 V thanks to greater physical insight and the adoption of trap-healing techniques. On the other hand, different technology platforms were employed to implement ADCs, and an ADC printed on foil was demonstrated for the first time. At the state of the art, ADCs on foil have reached an *SNDR* as high as 50 dB, equivalent to eight effective number of bits in 10 Hz bandwidth.

This concluding chapter aims at capturing the sense of these developments in relation to the fundamental objective of the research area as a whole, that of enabling analogue electronics on foil that meets the requirements of sensorisation. In view of the recent breakthroughs, flexible analogue electronics already provides a wealth of opportunities for a range of applications, as discussed in Section 5.1. Future developments, however, also present challenges, which demand further research efforts. A discussion of these is presented in Section 5.2.

5.1   *Immediate Opportunities for Flexible Analogue Electronics*

Recent developments evidence that exciting times are ahead for flexible analogue electronics. Firstly, circuit performance has reached levels suitable for a number of applications. Furthermore, flexible analogue circuits are now ready to be integrated with digital processing for the development of smart-sensor systems, and with printed batteries or compact energy harvesters to achieve autonomous operation. These immediate opportunities are discussed in the rest of this section.

5.1.1 Opportunities for Sensing Applications

The gain figures achieved by flexible analogue amplifiers in recent years indicate their suitability for the amplification of signals transduced by a wide range of sensors. On top of the basic configuration with a differential amplifier on foil providing the required signal amplification (i.e., amplifier in open loop), recent breakthroughs attaining gain figures in the 40 − 60 dB range indicate the viability of amplification with negative feedback. The latter approach ensures a more robust control of the overall circuit performance and lower sensitivity to process parameter variations. Differential amplification with negative feedback has already been demonstrated in combination with a flexible photodetector [124], and, in principle, it could be extended seamlessly to a broad range of sensing elements – for instance, thermal, humidity, pressure, biological and gas sensors.

In terms of bandwidth, state-of-the-art amplifiers on foil comfortably reach the kilohertz region. Such bandwidth is more than sufficient for a large number of applications that involve human–machine interaction (e.g., wearable electronics), tracking of environmental parameters (e.g., smart homes and smart cities) and biological monitoring (e.g., for healthcare purposes).

When it comes to analogue-to-digital converters, it has been shown recently that an *SNDR* of 50 dB, equivalent to an effective number of bits equal to eight, is reachable in 10 bandwidth. As long as the front-end amplifiers provide the same level of dynamic range, this specification could enable a wide range of sensor-based applications. For instance, a minimum resolution of 8 bits is required for monitoring food quality during conservation, based on measurement of temperature, pH, humidity, etc. The same holds for smart pressure-sensor matrices aiming to detect a wide range of goods on a shelf, from metallic materials to potato chips. These are examples of typical low-cost/large-area applications that cannot be served by mainstream silicon technologies.

It is noteworthy that the opportunities opened up by flexible analogue electronics are not simply defined by performance levels, but also by the inherent lightness and flexibility. Therefore, the applications discussed previously may be pursued in unique form factors, enabling functionalities that are yet to be explored.

5.1.2   Integration with Digital Processing

The promising performance levels recently attained by flexible analogue electronics suggest that times are now ripe for its integration with digital processing. Such integration is key to the realisation of smart sensors.

Integration with digital processing may rely on organic and/or metal-oxide digital circuits on foil, or, alternatively, on conventional-digital silicon electronics. The former approach would lead to inherently flexible smart sensors, and would potentially enable the use of the same technological platform for the fabrication of both analogue and digital circuits. In spite of the great progress made over the past decade, organic and/or metal-oxide digital circuits can only provide signal processing with limited computational capabilities, incomparable to those offered by conventional silicon ICs. If sufficiently small, a silicon IC would not have a major impact on mechanical flexibility and cost. Ultimately, the choice of integrating flexible analogue electronics with an organic and/or metal-oxide digital processor instead of a digital silicon chip is a matter of convenience: the former option would lead to a more scalable integration and potentially lower cost, while the latter could afford greater signal processing complexity and is more realistic in the near future.

5.1.3   Integration with Printed Batteries and Compact Energy Harvesters

The low operational voltage (5 V) and extremely limited power demand ($\approx 10 \ \mu W$) of state-of-the-art flexible analogue electronics point towards its integration with printed batteries and compact energy harvesters. This is key to realising flexible self-powered smart sensors. For instance, commercially available printed batteries (with thickness in the region of 500 μm, thus also capable of the desired flexibility) can already be configured to deliver the 5 V power supply voltage currently required by organic and/or metal-oxide differential amplifiers, while their capacity rating could ensure a lifetime of ≈ 30,000 h. Alternatively, self-powered operation could be realised by a range of flexible energy harvesters that capture energy from the environment (e.g., flexible solar cells, radiofrequency harvesting circuits). Compared with the battery-based approach, energy harvesters bring along the need of turning a low-voltage energy source into the desired power supply voltage (via boost converter circuits). At the same time, they provide the benefit of a virtually limitless lifetime.

5.2 Outlook and Challenges

While flexible analogue electronics already provides a wealth of opportunities, we believe that its full potential is yet to be unravelled. Specifically, this concerns performance, variability and operational voltage, all of which constitute key research areas that demand further investigation.

5.2.1 Circuit Performance

Recent developments in flexible amplifiers show that complementary integration is key to achieving large gain and bandwidth. Therefore, ongoing progress in n-channel OSs and p-channel oxide semiconductors is of particular interest for flexible analogue electronics. This is in view of the currently lower performance and stability of these semiconductors with respect to their homologous counterparts, and their relevance to complementary integration in an organic-only or metal-oxide-only framework. Hybrid technologies, instead, represent a reasonably well-established platform for high-performance complementary circuits on foil. Hybrid organic/metal-oxide technologies would also benefit from further material developments: on one hand, the reduction of the processing temperature of solution-based metal-oxide semiconductors would enable more facile integration on plastic substrates; on the other hand, the development of solution-processible p-channel OSs with higher mobility would enable the full exploitation of the bandwidth potential of the n-channel metal-oxide counterpart. These material developments demand combined advances in chemical synthesis and physical insight.

Large gain figures are essential to access the versatility and robustness of negative-feedback-loop circuit configurations. The gain of complementary circuits is largely dependent on the semi-conductors involved, through the transconductance and the output resistance of the corresponding transistors. Thus far, very limited insight is available on the maximum gain that can be achieved from organic and metal-oxide semiconductors. Indeed, out of convenience, device models have been largely borrowed from silicon technology. This is, however, a limiting factor, as the operation of organic and metal-oxide transistors can deviate substantially from such models. While modelling of mobility – which determines transconductance – is generally well developed, a specific aspect that urgently needs in-depth investigation is the output resistance of organic and metal-oxide transistors, in particular in relation to material and device parameters. Other crucial points where modelling should improve are contact effects and subthreshold behaviour, both of which relate significantly to transistor gain.

Large bandwidth becomes necessary in the presence of fast signal sources. As bandwidth is closely related to mobility, the push for improved bandwidth figures requires the development of organic and metal-oxide semiconductors with higher mobility. Therefore, future developments of higher-bandwidth flexible analogue electronics require a close monitoring of the synthetic achievements in this direction.

The improvements that a complementary technology can bring to the performance of analogue and digital building blocks would also directly affect ADC performance. A higher gain-bandwidth in amplifiers translates into higher accuracy in many ADC architectures. Furthermore, the availability of both p- and n-channel transistors broadens the range of architecture choices and analogue circuit techniques available, allowing for a lower power consumption at the same level of resolution.

5.2.2 Variability

The variability of characteristic device parameters in organic and metal-oxide semiconductor technologies can impose significant limitations on performance and functionality (cf. discussion on offset voltage and *CMRR* of differential amplifiers in Section 3.3, and on nonlinearity in ADC characteristics in Section 4.2). Therefore, further progress demands addressing this issue at all relevant levels: materials, processing and circuit design.

Acting on the causes underlying variability is an essential objective in its own right. Device variability involves semiconductor non-uniformity in microstructure, and trapping effects manifesting at varying levels. Counteracting non-uniformity in microstructure may involve the engineering of deposition conditions of materials that are polycrystalline. A superior approach is found, however, in the adoption of semiconductors that do not present long-range structural order, namely near-amorphous semiconductors. This is certainly the case of ternary and quaternary metal-oxide semiconductors. In the organic semiconductor domain, instead, this approach points to amorphous polymers. The development of high-performance amorphous polymers is thus an area of great interest for the reduction of circuit variability. Finally, trapping effects manifesting to a varying extent across batches or even within a foil may constitute an important cause of variability, especially in low-voltage devices and circuits. Therefore, the development of strategies for trap passivation may be key to realising devices with narrow parameter distribution.

Besides acting on the causes underlying variability, it is important to pursue strategies that minimise the impact of variability on circuit performance. At a material level, an ideal organic, metal-oxide or hybrid organic/metal-oxide semiconductor technology is one allowing complementary integration, hence leading to an inherently smaller sensitivity of circuit performance to semiconductor parameter variations. Consequently, flexible organic electronics looks with great interest at synthetic developments of higher performance n-channel OSs and p-channel AMOxSs, as demanded by an organic-only or metal-oxide-only complementary framework. Similar considerations apply to synthetic efforts concerning semiconductors relevant to hybrid organic/metal-oxide complementary technologies.

Suppressing the sensitivity to variability by adopting a complementary technology is accompanied by the challenge of establishing complementary integration schemes that do not compound variability effects. The most promising approaches are those that involve minimal processing. Indeed, each additional process step in a given integration platform generally leads to greater variability. Minimal processing within a complementary framework is inherently allowed by additive methods such as on-demand deposition (i.e., inkjet printing), a strategy that obviously involves solution-processible semiconductors.

At a circuit design level, the adoption of negative feedback constitutes an attractive route, as it is known to desensitise circuit performance from parameter variations. The desensitisation effect is proportional to the amplification achieved by the gain block in the feedback loop, however. This implies that amplifiers with large gains are required, preferably greater than 60 dB. Therefore, besides being key to greater performance, the pursuit of higher amplification gains constitutes a worthwhile goal in order to limit the impact of process parameter variations.

5.2.3 Low-Voltage Operation and Low Power Dissipation

An essential element to ensure the success of analogue circuits on foil is their ability to operate within a limited supply voltage and power consumption, especially in wirelessly powered and battery-powered applications. This demands circuits that can function with a power supply voltage not greater than 5 V, and preferably below 3 V.

While the 5 V threshold is within their capability, unipolar metal-oxide semiconductor technologies operating at below 3 V are yet to be demonstrated. This may be well within reach, considering the level of control on the switch-on characteristics achieved, for instance, by sputter-coated metal-oxide transistors.

Low-voltage operation is a much greater challenge for solution-processed technologies. Indeed, the majority of flexible analogue electronics works to date report power supply values well in excess of 5 V, and most generally in the 20 – 60 V region. The challenge is particularly significant for complementary circuits, in view of the complexity of integrating two semiconductors of opposite polarity both capable of low-voltage operation. A recent breakthrough has shown that combining a hybrid platform with a trap-healing strategy (for the p-channel organic semiconductor) can deliver complementary differential amplifiers with cutting-edge performance at a power supply of 5 V. While this result shows a way forward, substantial efforts are required to achieve complex complementary circuits functioning at even lower power supply values ($\approx$ 3 V).

The reduction of the operational voltage of complementary analogue circuits demands greater physical insight into low-voltage thin-film transistor operation, in particular with respect to threshold voltage. Indeed, while the subthreshold slope has been generally traced to disorder effects, and strategies to prevent the latter have been successfully devised, threshold voltage has been conventionally regarded as a given of a specific material/technology. This is in contrast to the precise understanding and control of threshold voltage in silicon technology. It is thus of paramount importance to gain greater insight into the origin of threshold voltage in solution-processed transistors, and to develop a systematic approach for its control and reduction within a circuit integration platform.

A further approach to low-voltage flexible analogue circuits yet to be fully explored is that of transistor operation in subthreshold. In this biasing region, the transistor transconductance is typically maximum for any given drain-source current (maximum g_m/I_{DS}). Though a higher intrinsic gain can be achieved, the bandwidth is extremely limited, due to the low current levels attainable. Nevertheless, this approach can be favourable in certain applications for which speed is not a determinative requirement, while extremely high power efficiency is demanded.

5.3　Concluding Remarks

This chapter has captured the ambitions of flexible analogue electronics. Performance levels are now ripe for the exploration of a wealth of self-powered sensing applications, e.g., in wearable electronics, environmental monitoring for smart homes and smart cities and biological monitoring for healthcare. Flexible analogue electronics is still in its youth, however, and its full potential is yet to be unravelled. Further research efforts – at the synthetic, device-physics, process-integration and circuit-design levels – are therefore needed to achieve performance maturity, which would enable flexible analogue electronics to play a central role in the sensorisation revolution of our time.

Acronyms

ADC	analogue-to-digital converter
ALD	atomic layer deposition
AMOxS	amorphous metal-oxide semiconductor
ASDM	asynchronous sigma-delta modulator
a-Si	amorphous silicon
BG	bottom gate
BGE	bootstrapped gain enhancement
BW	bandwidth
CMRR	common-mode rejection ratio
CVD	chemical vapour deposition
DAC	digital-to-analogue converter
DC	direct current
DNL	differential nonlinearity
ENOB	effective number of bits
FS	full scale
HOMO	highest occupied molecular orbital
IC	integrated circuit
IGZO	indium gallium zinc oxide
INL	integral nonlinearity
IZO	indium zinc oxide
LSB	least significant bit
LUMO	lowest unoccupied molecular orbital
MOSFET	metal-oxide-semiconductor field-effect transistor
OS	organic semiconductor
OSR	oversampling ratio
P4VP	poly(4-vinyl phenol)
PECVD	plasma-enhanced chemical vapour deposition
PEN	polyethylene naphthalate
PET	polyethylene terephthalate
PI	polyimide
PMMA	poly(methyl methacrylate)
poly-Si	polycrystalline silicon

PS	polystyrene
PWM	pulse width modulated
R2R	roll-to-roll
RFID	radio-frequency identification
RMS	root mean square
RT	room temperature
SAR	successive approximation
SDM	sigma-delta modulator
SFDR	spurious-free dynamic range
SNDR	signal-to-noise-and-distortion ratio
SNR	signal-to-noise ratio
TFT	thin-film transistor
TG	top gate
VCO	voltage-controlled oscillator
VTC	voltage transfer characteristic

References

[1] M. G. Kane, J. Campi, M. S. Hammond, F. P. Cuomo, B. Greening, C. D. Sheraw, J. A. Nichols, D. J. Gundlach, J. R. Huang, C. C. Kuo, L. Jia, H. Klauk and T. N. Jackson, 'Analog and digital circuits using organic thin-film transistors on polyester substrates', *IEEE Electron Device Lett.*, vol. 21, no. 11, pp. 534–536, 2000.

[2] N. Gay, W. J. Fischer, M. Halik, H. Klauk, U. Zschieschang and G. Schmid, 'Analog signal processing with organic FETs', in *2006 IEEE Int. Solid-State Circuits Conf.*, San Francisco, CA, USA, 2006, pp. 1070–1079.

[3] H. Marien, M. Steyaert, E. van Veenendaal and P. Heremans, 'Analog techniques for reliable organic circuit design on foil applied to an 18 dB single-stage differential amplifier', *Org. Electron.*, vol. 11, no. 8, pp. 1357–1362, 2010.

[4] H. Marien, M. S. J. Steyaert, E. van Veenendaal and P. Heremans, 'Analog building blocks for organic smart sensor systems in organic thin-film transistor technology on flexible plastic foil', *IEEE J. Solid-State Circuits*, vol. 47, no. 7, pp. 1712–1720, 2012.

[5] H. Marien, M. Steyaert, N. van Aerle and P. Heremans, 'An analog organic first-order CT DELTA SIGMA ADC on a flexible plastic substrate with 26.5dB precision', in *2010 IEEE Int. Solid-State Circuits Conf.*, San Francisco, CA, USA, 2010, pp. 136–137.

[6] Y. H. Tai, H. L. Chiu, L. S. Chou and C. H. Chang, 'Boosted gain of the differential amplifier using the second gate of the dual-gate a-IGZO TFTs', *IEEE Electron Device Lett.*, vol. 33, no. 12, pp. 1729–1731, 2012.

[7] G. Maiellaro, E. Ragonese, A. Castorina, S. Jacob, M. Benwadih, R. Coppard and E. Cantatore, 'High-gain operational transconductance amplifiers in a printed complementary organic TFT technology on flexible foil', *IEEE Trans. Circuits Syst.*, vol. 60, no. 12, pp. 3117–3125, 2013.

[8] V. Pecunia, M. Nikolka, A. Sou, I. Nasrallah, A. Y. Amin, I. McCulloch and H. Sirringhaus, 'Trap healing for high-performance low-voltage

polymer transistors and solution-based analog amplifiers on foil', *Adv. Mater.*, vol. 29, no. 23, p. 1606938, Jun. 2017.

[9] C. Garripoli, J.-L. P. J. van der Steen, E. Smits, G. H. Gelinck, A. H. M. van Roermund and E. Cantatore, '15.3 An a-IGZO asynchronous delta-sigma modulator on foil achieving up to 43dB SNR and 40dB SNDR in 300Hz bandwidth', in *2017 IEEE* Int. *Solid-State Circuits Conf.*, San Francisco, CA, USA, 2017, pp. 260–261.

[10] M. Marien, M. Steyaert and P. Heremans, *Analog Organic Electronics*. New York, NY: Springer, 2013.

[11] D. Raiteri, E. Cantatore and A. H. M. van Roermund, *Circuit Design on Plastic Foils*. Cham: Springer International Publishing, 2015.

[12] S. Abdinia, A. van Roermund and E. Cantatore, *Design of Organic Complementary Circuits and Systems on Foil*. Cham: Springer International Publishing, 2015.

[13] H. Sun, Y. Xu and Y. Noh, 'Flexible organic amplifiers', *IEEE Trans. Electron Devices*, vol. 64, pp. 1944–1954, 2017.

[14] J. H. Huijsing, 'Smart sensor systems: Why? Where? How?', in *Smart Sensor Systems*, G. C. M. Meijer, Ed. Chichester, UK: John Wiley & Sons, Ltd, 2008, pp. 1–21.

[15] H. Geng, Ed., *Internet of Things and Data Analytics Handbook*. Hoboken, NJ, USA: John Wiley & Sons, Inc., 2017.

[16] H. Chaouchi, 'Introduction to the Internet of Things', in *The Internet of Things*, H. Chaouchi, Ed. London, UK: ISTE Ltd and Hoboken, NJ, USA: John Wiley & Sons, Inc., 2010, pp. 1–33.

[17] R. D. Jansen-van Vuuren, A. Armin, A. K. Pandey, P. L. Burn and P. Meredith, 'Organic photodiodes: The future of full color detection and image sensing', *Adv. Mater.*, vol. 28, no. 24, pp. 4766–4802, Jun. 2016.

[18] A. Pierre and A. C. Arias, 'Solution-processed image sensors on flexible substrates', *Flex. Print. Electron.*, vol. 1, no. 4, p.43001, Dec. 2016.

[19] G. Pace, A. Grimoldi, M. Sampietro, D. Natali and M. Caironi, 'Printed photodetectors', *Semicond. Sci. Technol.*, vol. 30, no. 10, p. 104006, Oct. 2015.

[20] T. Li, L. Li, H. Sun, Y. Xu, X. Wang, H. Luo, Z. Liu and T. Zhang, 'Porous ionic membrane based flexible humidity sensor and its multifunctional applications', *Adv. Sci.*, vol. 4, no. 5, p. 1600404, May 2017.

[21] S. Borini, R. White, D. Wei, M. Astley, S. Haque, E. Spigone, N. Harris, J. Kivioja and T. Ryhänen, 'Ultrafast graphene oxide humidity sensors', *ACS Nano*, vol. 7, no. 12, pp. 11166–11173, 2013.

[22] S. Kano, K. Kim and M. Fujii, 'Fast-response and flexible nanocrystal-based humidity sensor for monitoring human respiration and water evaporation on skin', *ACS Sensors*, vol. 2, no. 6, pp. 828–833, Jun. 2017.

[23] T. Yokota, Y. Inoue, Y. Terakawa, J. Reeder, M. Kaltenbrunner, T. Ware, K. Yang, K. Mabuchi, T. Murakawa, M. Sekino, W. Voit, T. Sekitani and T. Someya, 'Ultraflexible, large-area, physiological temperature sensors for multipoint measurements', *Proc. Natl. Acad. Sci.*, vol. 112, no. 47, pp. 14533–14538, Nov. 2015.

[24] X. Ren, K. Pei, B. Peng, Z. Zhang, Z. Wang, X. Wang and P. K. L. Chan, 'A low-operating-power and flexible active-matrix organic-transistor temperature-sensor array', *Adv. Mater.*, vol. 28, no. 24, pp. 4832–4838, Jun. 2016.

[25] T. Q. Trung and N.-E. Lee, 'Flexible and stretchable physical sensor integrated platforms for wearable human-activity monitoring and personal healthcare', *Adv. Mater.*, vol. 28, no. 22, pp. 4338–4372, Jun. 2016.

[26] C. Yan, J. Wang and P. S. Lee, 'Stretchable graphene thermistor with tunable thermal index', *ACS Nano*, vol. 9, no. 2, pp. 2130–2137, Feb. 2015.

[27] M. Kaltenbrunner, T. Sekitani, J. Reeder, T. Yokota, K. Kuribara, T. Tokuhara, M. Drack, R. Schwödiauer, I. Graz, S. Bauer-Gogonea, S. Bauer and T. Someya, 'An ultra-lightweight design for imperceptible plastic electronics', *Nature*, vol. 499, no. 7459, pp.458–463, 2013.

[28] G. Schwartz, B. C.-K. Tee, J. Mei, A. L. Appleton, D. H. Kim, H. Wang and Z. Bao, 'Flexible polymer transistors with high pressure sensitivity for application in electronic skin and health monitoring', *Nat. Commun.*, vol. 4, no. 1859, May 2013.

[29] W. Lee, H. Koo, J. Sun, J. Noh, K.-S. Kwon, C. Yeom, Y. Choi, K. Chen, A. Javey and G. Cho, 'A fully roll-to-roll gravure-printed carbon nanotube-based active matrix for multi-touch sensors', *Sci. Rep.*, vol. 5, p. 17707, Dec. 2015.

[30] C. Zhang, P. Chen and W. Hu, 'Organic field-effect transistor-based gas sensors', *Chem. Soc. Rev.*, vol. 44, no. 8, pp. 2087–2107, 2015.

[31] T. Wang, D. Huang, Z. Yang, S. Xu, G. He, X. Li, N. Hu, G. Yin, D. He and L. Zhang, 'A review on graphene-based gas/vapor sensors with

unique properties and potential applications', *Nano-Micro Lett.*, vol. 8, no. 2, pp. 95–119, 2016.

[32] T. Wang, Y. Guo, P. Wan, H. Zhang, X. Chen and X. Sun, 'Flexible transparent electronic gas sensors', *Small*, vol. 12, no. 28, pp. 3748–3756, 2016.

[33] Y. H. Kim, S. J. Kim, Y.-J. Kim, Y.-S. Shim, S. Y. Kim, B. H. Hong and H. W. Jang, 'Self-activated transparent all-graphene gas sensor with endurance to humidity and mechanical bending', *ACS Nano*, vol. 9, no. 10, pp. 10453–10460, Oct. 2015.

[34] M. Magliulo, M. Y. Mulla, M. Singh, E. Macchia, A. Tiwari, L. Torsi and K. Manoli, 'Printable and flexible electronics: From TFTs to bioelectronic devices', *J. Mater. Chem. C*, vol. 3, no. 48, pp. 12347–12363, 2015.

[35] C. Liao, M. Zhang, M. Y. Yao, T. Hua, L. Li and F. Yan, 'Flexible organic electronics in biology: Materials and devices', *Adv. Mater.*, vol. 27, no. 46, pp. 7493–7527, Dec. 2015.

[36] D. Khodagholy, J. N. Gelinas, T. Thesen, W. Doyle, O. Devinsky, G. G. Malliaras and G. Buzsáki, 'NeuroGrid: Recording action potentials from the surface of the brain', *Nat. Neurosci.*, vol. 18, no. 2, pp. 310–315, Dec. 2014.

[37] C. M. Lochner, Y. Khan, A. Pierre and A. C. Arias, 'All-organic optoelectronic sensor for pulse oximetry', *Nat. Commun.*, vol. 5, no. 5745, Dec. 2014.

[38] T. Takahashi, Z. Yu, K. Chen, D. Kiriya, C. Wang, K. Takei, H. Shiraki, T. Chen, B. Ma and A. Javey, 'Carbon nanotube active-matrix backplanes for mechanically flexible visible light and X-ray imagers', *Nano Lett.*, vol. 13, no. 11, pp. 5425–5430, Nov. 2013.

[39] G. Horowitz, R. Hajlaoui, R. Bourguiga and M. Hajlaoui, 'Theory of the organic field-effect transistor', *Synth. Met.*, vol. 101, no. 1–3, pp. 401–404, May 1999.

[40] O. Marinov, M. J. Deen, U. Zschieschang and H. Klauk, 'Organic thin-film transistors: Part I-compact DC modeling', *IEEE Trans. Electron Devices*, vol. 56, no. 12, pp. 2952–2961, 2009.

[41] P. Servati, D. Striakhilev and A. Nathan, 'Above-threshold parameter extraction and modeling for amorphous silicon thin-film transistors', *IEEE Trans. Electron Devices*, vol. 50, no. 11, pp. 2227–2235, Nov. 2003.

[42] P. Heremans, A. K. Tripathi, A. de Jamblinne de Meux, E. C. P. Smits, B. Hou, G. Pourtois and G. H. Gelinck, 'Mechanical and electronic properties of thin-film transistors on plastic, and their integration in flexible electronic applications', *Adv. Mater.*, vol. 28, no. 22, pp. 4266–4282, 2016.

[43] H. Gleskova, S. Wagner and Z. Suo, 'Failure resistance of amorphous silicon transistors under extreme in-plane strain', *Appl. Phys. Lett.*, vol. 75, no. 19, pp. 3011–3013, Nov. 1999.

[44] F. Silveira, D. Flandre and P. G. A. Jespers, 'A gm/ID based methodology for the design of CMOS analog circuits and its application to the synthesis of a silicon-on-insulator micropower OTA', *IEEE J. Solid-State Circuits*, vol. 31, no. 9, pp. 1314–1319, 1996.

[45] B. Murmann and W. X. W. Xiong, 'Design of analog circuits using organic field-effect transistors', in *2010 IEEE/ACM Int. Conf. Comput.-Aided Des.*, San Jose, CA, USA, 2010, pp. 504–507.

[46] H. Sirringhaus, '25th anniversary article: Organic field-effect transistors: The path beyond amorphous silicon', *Adv. Mater.*, vol. 26, no. 9, pp. 1319–1335, Mar. 2014.

[47] X. Yu, T. J. Marks and A. Facchetti, 'Metal oxides for optoelectronic applications', *Nat. Mater.*, vol. 15, no. 4, pp. 383–396, 2016.

[48] J. Yeon Kwon and J. Kyeong Jeong, 'Recent progress in high performance and reliable n-type transition metal oxide-based thin film transistors', *Semicond. Sci. Technol.*, vol. 30, no. 2, p. 24002, 2015.

[49] Y. Zhao, Y. Guo and Y. Liu, '25th anniversary article: Recent advances in n-type and ambipolar organic field-effect transistors', *Adv. Mater.*, vol. 25, no. 38, pp. 5372–5391, Oct. 2013.

[50] H. Sirringhaus, P. J. Brown, R. H. Friend, M. M. Nielsen, K. Bechgaard, B. M. W. Langeveld-Voss, A. J. H. Spiering, R. A. J. Janssen, E. W. Meijer, P. Herwig and D. M. de Leeuw, 'Two-dimensional charge transport in self-organized, high-mobility conjugated polymers', *Nature*, vol. 401, no. 6754, pp. 685–688, 1999.

[51] I. McCulloch, M. Heeney, C. Bailey, K. Genevicius, I. Macdonald, M. Shkunov, D. Sparrowe, S. Tierney, R. Wagner, W. Zhang, M. L. Chabinyc, R. J. Kline, M. D. McGehee and M. F. Toney, 'Liquid-crystalline semiconducting polymers with high charge-carrier mobility', *Nat. Mater.*, vol. 5, no. 4, pp. 328–333, Apr. 2006.

[52] C. B. Nielsen, M. Turbiez and I. McCulloch, 'Recent advances in the development of semiconducting DPP-containing polymers for transistor applications', *Adv. Mater.*, vol. 25, no. 13, pp. 1859–1880, 2013.

[53] G. Kim, S. J. Kang, G. K. Dutta, Y. K. Han, T. J. Shin, Y. Y. Noh and C. Yang, 'A thienoisoindigo-naphthalene polymer with ultrahigh mobility of 14.4 cm2/Vs that substantially exceeds benchmark values for amorphous silicon semiconductors', *J. Am. Chem. Soc.*, vol. 136, no. 26, pp. 9477-9483, 2014.

[54] Y. Zhao, Y. Guo and Y. Liu, '25th anniversary article: Recent advances in n-type and ambipolar organic field-effect transistors', *Adv. Mater.*, vol. 25, no. 38, pp. 5372-5391, 2013.

[55] J. Rivnay, M. F. Toney, Y. Zheng, I. V. Kauvar, Z. Chen, V. Wagner, A. Facchetti and A. Salleo, 'Unconventional face-on texture and exceptional in-plane order of a high mobility n-type polymer', *Adv. Mater.*, vol. 22, no. 39, pp. 4359-4363, 2010.

[56] H. Hosono, 'Ionic amorphous oxide semiconductors: Material design, carrier transport, and device application', *J. Non. Cryst. Solids*, vol. 352, no. 9-20, pp. 851-858, Jun. 2006.

[57] T. Kamiya, K. Nomura and H. Hosono, 'Electronic structure of the amorphous oxide semiconductor a-InGaZnO4-x : Tauc-Lorentz optical model and origins of subgap states', *Phys. Status Solidi*, vol. 206, no. 5, pp. 860-867, May 2009.

[58] Z. Wang, P. K. Nayak, J. A. Caraveo-Frescas and H. N. Alshareef, 'Recent developments in p-type oxide semiconductor materials and devices', *Adv. Mater.*, vol. 28, no. 20, pp. 3831-3892, May 2016.

[59] K. Nomura, H. Ohta, A. Takagi, T. Kamiya, M. Hirano and H. Hosono, 'Room-temperature fabrication of transparent flexible thin-film transistors using amorphous oxide semiconductors', *Nature*, vol. 432, no. 7016, pp. 488-492, Nov. 2004.

[60] S. D. Brotherton, *Introduction to Thin Film Transistors*. Cham, Switzerland: Springer, 2013.

[61] M. Nikolka, I. Nasrallah, B. Rose, M. K. Ravva, K. Broch, A. Sadhanala, D. Harkin, J. Charmet, M. Hurhangee, A. Brown, S. Illig, P. Too, J. Jongman, I. McCulloch, J.-L. Bredas and H. Sirringhaus, 'High operational and environmental stability of high-mobility conjugated polymer field-effect transistors through the use of molecular additives', *Nat. Mater.*, vol. 16, no. 3, pp. 356-362, 2017.

[62] Y. H. Tai, C. Y. Chang, C. L. Hsieh, Y. H. Yang, W. K. Chao and H. E. Chen, 'Dependence of the noise behavior on the drain current for thin film transistors', *IEEE Electron Device Lett.*, vol. 35, no. 2, pp. 229-231, 2014.

[63] L.-Y. Su, H.-K. Lin, C.-C. Hung and J. Huang, 'Role of HfO2/SiO2 gate dielectric on the reduction of low-frequent noise and the enhancement of a-IGZO TFT electrical performance', *J. Disp. Technol.*, vol. 8, no. 12, pp. 695–698, Dec. 2012.

[64] T.-C. Fung, G. Baek and J. Kanicki, 'Low frequency noise in long channel amorphous In–Ga–Zn–O thin film transistors', *J. Appl. Phys.*, vol. 108, no. 7, p. 74518, Oct. 2010.

[65] O. D. Jurchescu, B. H. Hamadani, H. D. Xiong, S. K. Park, S. Subramanian, N. M. Zimmerman, J. E. Anthony, T. N. Jackson and D. J. Gundlach, 'Correlation between microstructure, electronic properties and flicker noise in organic thin film transistors', *Appl. Phys. Lett.*, vol. 92, no. 13, pp. 1–4, 2008.

[66] Z. Jia, I. Meric, K. L. Shepard and I. Kymissis, 'Doping and illumination dependence of 1/f noise in pentacene thin-film transistors', *IEEE Electron Device Lett.*, vol. 31, no. 9, pp. 1050–1052, 2010.

[67] L. F. Deng, Y. R. Liu, H. W. Choi, C. M. Che and P. T. Lai, 'Improved performance of pentacene OTFTs with HfLaO gate dielectric by using fluorination and nitridation', *IEEE Trans. Device Mater. Reliab.*, vol. 12, no. 2, pp. 520–528, 2012.

[68] R. A. Street, Ed., *Technology and Applications of Amorphous Silicon*, vol. 37. Berlin, Heidelberg: Springer Berlin Heidelberg, 2000.

[69] J. F. Wager, B. Yeh, R. L. Hoffman and D. A. Keszler, 'An amorphous oxide semiconductor thin-film transistor route to oxide electronics', *Curr. Opin. Solid State Mater. Sci.*, vol. 18, no. 2, pp. 53–61, 2014.

[70] B. Kheradmand-Boroujeni, G. C. Schmidt, D. Höft, R. Shabanpour, C. Perumal, T. Meister, K. Ishida, C. Carta, A. C. Hübler and F. Ellinger, 'Analog characteristics of fully printed flexible organic transistors fabricated with low-cost mass-printing techniques', *IEEE Trans. Electron Devices*, vol. 61, no. 5, pp. 1423–1430, 2014.

[71] B. Kheradmand-Boroujeni, G. C. Schmidt, D. Höft, K. Haase, M. Bellmann, K. Ishida, R. Shabanpour, T. Meister, C. Carta, A. C. Hübler and F. Ellinger, 'Small-signal characteristics of fully-printed high-current flexible all-polymer three-layer-dielectric transistors', *Org. Electron.*, vol. 34, pp. 267–275, 2016.

[72] P. G. Bahubalindruni, A. Kiazadeh, A. Sacchetti, J. Martins, A. Rovisco, V. G. Tavares, R. Martins, E. Fortunato and P. Barquinha, 'Influence of channel length scaling on InGaZnO TFTs characteristics: Unity current-gain cutoff frequency, intrinsic

voltage-gain, and on-resistance', *J. Disp. Technol.*, vol. 12, no. 6, pp. 515–518, Jun. 2016.

[73] J. Pekarik, D. Greenberg, B. Jagannathan, R. Groves, J. R. Jones, R. Singh, A. Chinthakindi, X. Wang, M. Breitwisch, D. Coolbaugh, P. Cottrell, J. Florkey, G. Freeman and R. Krishnasamy, 'RFCMOS technology from 0.25µm to 65nm: The state of the art', in *IEEE 2004 Custom Integr. Circuits Conf.*, Orlando, FL, USA, 2004, pp. 217–224.

[74] J. Robertson and B. Falabretti, 'Band offsets of high K gate oxides on high mobility semiconductors', *Mater. Sci. Eng. B*, vol. 135, no. 3, pp. 267–271, Dec. 2006.

[75] F.-C. Chen, C.-W. Chu, J. He, Y. Yang and J.-L. Lin, 'Organic thin-film transistors with nanocomposite dielectric gate insulator', *Appl. Phys. Lett.*, vol. 85, no. 15, p. 3295, 2004.

[76] C. Jung, A. Maliakal, A. Sidorenko and T. Siegrist, 'Pentacene-based thin film transistors with titanium oxide-polystyrene/polystyrene insulator blends: High mobility on high K dielectric films', *Appl. Phys. Lett.*, vol. 90, no. 6, p. 62111, 2007.

[77] P. Kim, X.-H. Zhang, B. Domercq, S. C. Jones, P. J. Hotchkiss, S. R. Marder, B. Kippelen and J. W. Perry, 'Solution-processible high-permittivity nanocomposite gate insulators for organic field-effect transistors', *Appl. Phys. Lett.*, vol. 93, p. 13302, 2008.

[78] R. Schroeder, L. A. Majewski and M. Grell, 'High-performance organic transistors using solution-processed nanoparticle-filled high-k polymer gate insulators', *Adv. Mater.*, vol. 17, no. 12, pp. 1535–1539, Jun. 2005.

[79] Q. Chen, K. Ren, B. Chu, Y. Liu, Q. M. Zhang, V. Bobnar and A. Levstik, 'Relaxor ferroelectric polymers – fundamentals and applications', *Ferroelectrics*, vol. 354, no. 1, pp. 178–191, Aug. 2007.

[80] P. W. Peacock and J. Robertson, 'Band offsets and Schottky barrier heights of high dielectric constant oxides', *J. Appl. Phys.*, vol. 92, no. 8, p. 4712, 2002.

[81] V. Pecunia, K. Banger and H. Sirringhaus, 'High-performance solution-processed amorphous-oxide-semiconductor TFTs with organic polymeric gate dielectrics', *Adv. Electron. Mater.*, vol. 1, no. 1-2, Feb. 2015.

[82] L.-L. Chua, P. K. H. Ho, H. Sirringhaus and R. H. Friend, 'High-stability ultrathin spin-on benzocyclobutene gate dielectric for polymer field-effect transistors', *Appl. Phys. Lett.*, vol. 84, no. 17, p. 3400, 2004.

[83] Y. Noh and H. Sirringhaus, 'Ultra-thin polymer gate dielectrics for top-gate polymer field-effect transistors', *Org. Electron.*, vol. 10, no. 1, pp. 174–180, Feb. 2009.

[84] J. Li, D. Liu, Q. Miao and F. Yan, 'The application of a high-k polymer in flexible low-voltage organic thin-film transistors', *J. Mater. Chem.*, vol. 22, no. 31, p. 15998, 2012.

[85] J. H. Cho, J. Lee, Y. Xia, B. Kim, Y. He, M. J. Renn, T. P. Lodge and C. D. Frisbie, 'Printable ion-gel gate dielectrics for low-voltage polymer thin-film transistors on plastic', *Nat. Mater.*, vol. 7, no. 11, pp. 900–906, Nov. 2008.

[86] I.-T. Cho, J.-M. Lee, J.-H. Lee and H.-I. Kwon, 'Charge trapping and detrapping characteristics in amorphous InGaZnO TFTs under static and dynamic stresses', *Semicond. Sci. Technol.*, vol. 24, no. 1, p. 15013, Jan. 2009.

[87] M. Halik, H. Klauk, U. Zschieschang, G. Schmid, C. Dehm, M. Schütz, S. Maisch, F. Effenberger, M. Brunnbauer and F. Stellacci, 'Low-voltage organic transistors with an amorphous molecular gate dielectric', *Nature*, vol. 431, no. 7011, pp. 963–966, 2004.

[88] H. Klauk, U. Zschieschang, J. Pflaum and M. Halik, 'Ultralow-power organic complementary circuits', *Nature*, vol. 445, no. 7129, pp. 745–748, Feb. 2007.

[89] J. Veres, S. D. Ogier, S. W. Leeming, D. C. Cupertino and S. Mohialdin Khaffaf, 'Low-k insulators as the choice of dielectrics in organic field-effect transistors', *Adv. Funct. Mater.*, vol. 13, no. 3, pp. 199–204, Mar. 2003.

[90] I. N. Hulea, S. Fratini, H. Xie, C. L. Mulder, N. N. Iossad, G. Rastelli, S. Ciuchi and A. F. Morpurgo, 'Tunable Fröhlich polarons in organic single-crystal transistors', *Nat. Mater.*, vol. 5, no. 12, pp. 982–986, Dec. 2006.

[91] K. Fukuda, Y. Takeda, Y. Yoshimura, R. Shiwaku, L. T. Tran, T. Sekine, M. Mizukami, D. Kumaki and S. Tokito, 'Fully-printed high-performance organic thin-film transistors and circuitry on one-micron-thick polymer films', *Nat. Commun.*, vol. 5, no. 4147, Jun. 2014.

[92] R. A. Nawrocki, N. Matsuhisa, T. Yokota and T. Someya, '300-nm imperceptible, ultraflexible, and biocompatible e-skin fit with tactile sensors and organic transistors', *Adv. Electron. Mater.*, vol. 2, no. 4, p. 1500452, Apr. 2016.

[93] H.-H. Hsieh and C.-C. Wu, 'Amorphous ZnO transparent thin-film transistors fabricated by fully lithographic and etching processes', *Appl. Phys. Lett.*, vol. 91, no. 1, p. 13502, 2007.

[94] Y.-L. Wang, L. N. Covert, T. J. Anderson, W. Lim, J. Lin, S. J. Pearton, D. P. Norton, J. M. Zavada and F. Ren, 'RF characteristics of room-temperature-deposited, small gate dimension indium zinc oxide TFTs', *Electrochem. Solid-State Lett.*, vol. 11, no. 3, p. H60-H62, 2008.

[95] M.-J. Yu, Y.-H. Yeh, C.-C. Cheng, C.-Y. Lin, G.-T. Ho, B. C.-M. Lai, C.-M. Leu, T.-H. Hou and Y.-J. Chan, 'Amorphous InGaZnO thin-film transistors compatible with roll-to-roll fabrication at room temperature', *IEEE Electron Device Lett.*, vol. 33, no. 1, pp. 47–49, Jan. 2012.

[96] F. Zhou, H. P. Lin, L. Zhang, J. Li, X. W. Zhang, D. B. Yu, X. Y. Jiang and Z. L. Zhang, 'Top-gate amorphous IGZO thin-film transistors with a SiO buffer layer inserted between active channel layer and gate insulator', *Curr. Appl. Phys.*, vol. 12, no. 1, pp. 228–232, Jan. 2012.

[97] S.-H. K. Park, C.-S. Hwang, H. Y. Jeong, H. Y. Chu and K. I. Cho, 'Transparent ZnO-TFT arrays fabricated by atomic layer deposition', *Electrochem. Solid-State Lett.*, vol. 11, no. 1, p. H10-H14, 2008.

[98] P. Poodt, R. Knaapen, A. Illiberi, F. Roozeboom and A. van Asten, 'Low temperature and roll-to-roll spatial atomic layer deposition for flexible electronics', *J. Vac. Sci. Technol. A Vacuum, Surfaces, Film.*, vol. 30, no. 1, p. 01A142, 2012.

[99] S. E. Potts, H. B. Profijt, R. Roelofs and W. M. M. Kessels, 'Room-temperature ALD of metal oxide thin films by energy-enhanced ALD', *Chem. Vap. Depos.*, vol. 19, no. 4–6, pp. 125–133, Jun. 2013.

[100] J. Lin, 'Printing processes and equipments', in *Printed Electronics*, Singapore: John Wiley & Sons Singapore Pte. Ltd, 2016, pp. 106–144.

[101] S. Khan, L. Lorenzelli and R. S. Dahiya, 'Technologies for printing sensors and electronics over large flexible substrates: A review', *IEEE Sens. J.*, vol. 15, no. 6, pp. 3164–3185, 2015.

[102] S. E. Burns, P. Cain, J. Mills, J. Wang and H. Sirringhaus, 'Inkjet printing of polymer thin-film transistor circuits', *MRS Bull.*, vol. 28, no. 11, pp. 829–834, Nov. 2003.

[103] D. Raiteri, P. van Lieshout, A. van Roermund and E. Cantatore, 'Positive-feedback level shifter logic for large-area electronics',

IEEE J. Solid-State Circuits, vol. 49, no. 2, pp. 524–535, Feb. 2014.

[104] T. Arai and Y. Shiraishi, '56.1: Invited paper : Manufacturing issues for oxide TFT technologies for large-sized AMOLED displays', *SID Symp. Dig. Tech. Pap.*, vol. 43, no. 1, pp. 756–759, 2012.

[105] Complementary Organic Semiconductor and Metal Integrated Circuits (COSMIC). (2014). 'Project public final report'. Complementary Organic Semiconductor and Metal Integrated Circuits, GA-Nr. 247681. [Online]. Available: http://cordis.europa .eu/docs/projects/cnect/1/247681/080/deliverables/001- COSMICpublicreportV41.pdf

[106] S. Il Kim, J.-S. Park, C. J. Kim, J. C. Park, I. Song and Y. S. Park, 'High reliable and manufacturable gallium indium zinc oxide thin-film transistors using the double layers as an active layer', *J. Electrochem. Soc.*, vol. 156, no. 3, p. H184-H187, 2009.

[107] W. Xiong, Y. Guo, U. Zschieschang, H. Klauk and B. Murmann, 'A 3-V, 6-bit C-2C digital-to-analog converter using complementary organic thin-film transistors on glass', *IEEE J. Solid-State Circuits*, vol. 45, no. 7, pp. 1380–1388, Jul. 2010.

[108] K. Myny, P. van Lieshout, J. Genoe, W. Dehaene and P. Heremans, 'Accounting for variability in the design of circuits with organic thin-film transistors', *Org. Electron.*, vol. 15, no. 4, pp. 937–942, 2014.

[109] X. Zhang, T. Ge and J. S. Chang, 'Fully-additive printed electronics: Transistor model, process variation and fundamental circuit designs', *Org. Electron.*, vol. 26, pp. 371–379, 2015.

[110] S. Jacob, M. Benwadih, J. Bablet, I. Chartier, R. Gwoziecki, S. Abdinia, E. Cantatore, L. Maddiona, F. Tramontana, G. Maiellaro, L. Mariucci, G. Palmisano and R. Coppard, 'High performance printed N and P-type OTFTs for complementary circuits on plastic substrate', in *Eur. Solid-State Device Res. Conf.*, Bordeaux, France, 2012, pp. 173–176.

[111] S. Abdinia, T.-H. Ke, M. Ameys, J. Li, S. Steudel, J. L. Vandersteen, B. Cobb, F. Torricelli, A. van Roermund and E. Cantatore, 'Organic CMOS line drivers on foil', *J. Disp. Technol.*, vol. 11, no. 6, pp. 564–569, Jun. 2015.

[112] D. Tu, K. Takimiya, U. Zschieschang, H. Klauk and R. Forchheimer, 'Modeling of drain current mismatch in organic thin-film transistors', *J. Disp. Technol.*, vol. 11, no. 6, pp. 559–563, Jun. 2015.

[113] T. Kamiya, K. Nomura and H. Hosono, 'Present status of amorphous In-Ga-Zn-O thin-film transistors', *Sci. Technol. Adv. Mater.*, vol. 11, no. 4, p. 44305, Aug. 2010.

[114] C. Ha, H. Lee, J. Kwon, S. Seok, C.-I. Ryoo, K. Yun, B. Kim, W. Shin and S. Cha, '69.2: distinguished paper: High reliable a-IGZO TFTs with self-aligned coplanar structure for large-sized ultrahigh-definition OLED TV', *SID Symp. Dig. Tech. Pap.*, vol. 46, no. 1, pp. 1020–1022, Jun. 2015.

[115] X. Yu, J. Smith, N. Zhou, L. Zeng, P. Guo, Y. Xia, A. Alvarez, S. Aghion, H. Lin, J. Yu, R. P. H. Chang, M. J. Bedzyk, R. Ferragut, T. J. Marks and A. Facchetti, 'Spray-combustion synthesis: Efficient solution route to high-performance oxide transistors', *Proc. Natl. Acad. Sci. U. S. A.*, vol. 112, no. 11, pp. 3217–3222, 2015.

[116] Y. S. Rim, H. Chen, Y. Liu, S. H. Bae, H. J. Kim and Y. Yang, 'Direct light pattern integration of low-temperature solution-processed all-oxide flexible electronics', *ACS Nano*, vol. 8, no. 9, pp. 9680–9686, 2014.

[117] H. Marien, M. S. J. Steyaert, E. van Veenendaal and P. Heremans, 'A fully integrated DELTA SIGMA ADC in organic thin-film transistor technology on flexible plastic foil', *IEEE J. Solid-State Circuits*, vol. 46, no. 1, pp. 276–284, 2011.

[118] I. Nausieda, K. K. Ryu, D. Da He, A. I. Akinwande, V. Bulovic and C. G. Sodini, 'Mixed-signal organic integrated circuits in a fully photolithographic dual threshold voltage technology', *IEEE Trans. Electron Devices*, vol. 58, no. 3, pp. 865–873, 2011.

[119] N. Gay and W. J. Fischer, 'OFET-based analog circuits for microsystems and RFID-sensor transponders', in *Polytronic 2007 – 6th Int. IEEE Conf. Polym. Adhes. Microelectron. Photonics*, Odaiba, Japan, 2007 pp. 143–148, 2007.

[120] J. Chang, X. Zhang, T. Ge and J. Zhou, 'Fully printed electronics on flexible substrates: High gain amplifiers and DAC', *Org. Electron.*, vol. 15, no. 3, pp. 701–710, 2014.

[121] K. Fukuda, T. Minamiki, T. Minami, M. Watanabe, T. Fukuda, D. Kumaki and S. Tokito, 'Printed organic transistors with uniform electrical performance and their application to amplifiers in biosensors', *Adv. Electron. Mater.*, vol. 1, no. 7, p. 1400052, Jul. 2015.

[122] C. Zysset, N. Münzenrieder, L. Petti, L. Buthe, G. A. Salvatore and G. Tröster, 'IGZO TFT-based all-enhancement operational

amplifier bent to a radius of 5 mm', *IEEE Electron Device Lett.*, vol. 34, no. 11, pp. 1394–1396, Nov. 2013.

[123] G. A. Salvatore, N. Münzenrieder, T. Kinkeldei, L. Petti, C. Zysset, I. Strebel, L. Büthe and G. Tröster, 'Wafer-scale design of lightweight and transparent electronics that wraps around hairs', *Nat. Commun.*, vol. 5, p. 2982, Jan. 2014.

[124] G. Maiellaro, E. Ragonese, R. Gwoziecki, S. Jacobs, N. Marjanovic, M. Chrapa, J. Schleuniger and G. Palmisano, 'Ambient light organic sensor in a printed complementary organic TFT technology on flexible plastic foil', *IEEE Trans. Circuits Syst. I Regul. Pap.*, vol. 61, no. 4, pp.1036–1043, 2014.

[125] S. Abdinia, F. Torricelli, G. Maiellaro, R. Coppard, A. Daami, S. Jacob, L. Mariucci, G. Palmisano, E. Ragonese, F. Tramontana, A. H. M. van Roermund, and E. Cantatore, 'Variation-based design of an AM demodulator in a printed complementary organic technology', *Org. Electron.*, vol. 15, no. 4, pp. 904–912, 2014.

[126] V. Vaidya, D. M. Wilson, X. Zhang and B. Kippelen, 'An organic complementary differential amplifier for flexible AMOLED applications', in *2010 IEEE Int. Symp. Circuits Syst.*, Paris, France, 2010, pp. 3260–3263.

[127] M. Guerin, A. Daami, S. Jacob, E. Bergeret, E. Bènevent, P. Pannier and R. Coppard, 'High-gain fully printed organic complementary circuits on flexible plastic foils', *IEEE Trans. Electron Devices*, vol. 58, no. 10, pp. 3587–3593, 2011.

[128] K. Ishida, T. C. Huang, K. Honda, T. Sekitani, H. Nakajima, H. Maeda, M. Takamiya, T. Someya and T. Sakurai, 'A 100-V AC energy meter integrating 20-V organic CMOS digital and analog circuits with a floating gate for process variation compensation and a 100-v organic pMOS rectifier', *IEEE J. Solid-State Circuits*, vol. 47, no. 1, pp. 301–309, 2012.

[129] S. Lee, A. Reuveny, N. Matsuhisa, R. Nawrocki, T. Yokota and T. Someya, 'Enhancement of closed-loop gain of organic amplifiers using double-gate structures', *IEEE Electron Device Lett.*, vol. 37, no. 6, pp. 770–773, 2016.

[130] T.-C. Huang, K. Fukuda, C.-M. Lo, Y.-H. Yeh, T. Sekitani, T. Someya and K.-T. Cheng, 'Pseudo-CMOS: A design style for low-cost and robust flexible electronics', *IEEE Trans. Electron Devices*, vol. 58, no. 1, pp. 141–150, Jan. 2011.

[131] R. Shabanpour, T. Meister, K. Ishida, L. Petti, N. Münzenrieder, G. A. Salvatore, B. K. Boroujeni, C. Carta, G. Tröster and F. Ellinger, 'High gain amplifiers in flexible self-aligned a-IGZO thin-film-transistor technology', in *2014 21st IEEE Int. Conf. on Electron., Circuits and Syst.*, Marseille, France, 2014, pp 108–111.

[132] R. Shabanpour, K. Ishida, T. Meister, N. Münzenrieder, L. Petti, G. Salvatore, B. Kheradmand-Boroujeni, C. Carta, G. Tröster and F. Ellinger, 'A 70° phase margin OPAMP with positive feedback in flexible a-IGZO TFT technology', in *2015 IEEE 58th Int. Midwest Symp. Circuits Syst.*, Fort Collins, CO, USA, 2015, pp. 1–4.

[133] C. Garripoli, J.-L. P. J. van der Steen, F. Torricelli, M. Ghittorelli, G. H. Gelinck, A. H. M. van Roermund, and E. Cantatore, 'Analogue frontend amplifiers for bio-potential measurements manufactured with a-IGZO TFTs on flexible substrate', *IEEE J. Emerg. Sel. Top. Circuits Syst.*, Fort Collins, CO, USA, 2015, pp. 1–4.

[134] M. G. Kane, J. Campi, F. P. Cuomo and B. K. Greening, 'Fast organic circuits on flexible polymeric substrates', in *Int. Electron Devices Meeting 2000*, San Francisco, CA, USA, 2000, pp. 619–622.

[135] H. Sirringhaus, T. Kawase, R. H. Friend, T. Shimoda, M. Inbasekaran, W. Wu and E. P. Woo, 'High-resolution inkjet printing of all-polymer transistor circuits', *Science*, vol. 290, no. 5499, pp. 2123–2126, 2000.

[136] K. Myny, E. van Veenendaal, G. H. Gelinck, J. Genoe, W. Dehaene and P. Heremans, 'An 8-bit, 40-instructions-per-second organic microprocessor on plastic foil', *IEEE J. Solid-State Circuits*, vol. 47, no. 1, pp. 284–291, Jan. 2012.

[137] K. Myny, M. Rockelé, A. Chasin, D. V. Pham, J. Steiger, S. Botnaras, D. Weber, B. Herold, J. Ficker, B. der van Putten, G. H. Gelinck, J. Genoe, W. Dehaene and P. Heremans, 'Bidirectional communication in an HF hybrid organic/solution-processed metal-oxide RFID tag', *IEEE Trans. Electron Devices*, vol. 61, no. 7, pp. 2387–2393, Jul. 2014.

[138] A. Daami, C. Bory, M. Benwadih, S. Jacob, R. Gwoziecki, I. Chartier, R. Coppard, C. Serbutoviez, L. Maddiona, E. Fontana and A. Scuderi, 'Fully printed organic CMOS technology on plastic substrates for digital and analog applications', in *IEEE Int. Solid-State Circuits Conf.*, San Francisco, CA, USA, 2011, pp. 328–329.

[139] S. Jacob, S. Abdinia, M. Benwadih, J. Bablet, I. Chartier, R. Gwoziecki, E. Cantatore, A. H. M. van Roermund, L. Maddiona,

F. Tramontana, G. Maiellaro, L. Mariucci, M. Rapisarda, G. Palmisano and R. Coppard, 'High performance printed N and P-type OTFTs enabling digital and analog complementary circuits on flexible plastic substrate', *Solid. State. Electron.*, vol. 84, pp.167–178, 2013.

[140] K. Ishida, R. Shabanpour, B. K. Boroujeni, T. Meister, C. Carta, F. Ellinger, L. Petti, N. S. Münzenrieder, G. A. Salvatore and G. Tröster, '22.5 dB open-loop gain, 31 kHz GBW pseudo-CMOS based operational amplifier with a-IGZO TFTs on a flexible film', in *2014 IEEE Asian Solid-State Circuits Conf.*, Kaohsiung, Taiwan, 2014, pp. 313–316.

[141] W. L. Kalb, S. Haas, C. Krellner, T. Mathis and B. Batlogg, 'Trap density of states in small-molecule organic semiconductors: A quantitative comparison of thin-film transistors with single crystals', *Phys. Rev. B*, vol. 81, no. 15, pp. 1–13, 2010.

[142] P. J. Diemer, Z. A. Lamport, Y. Mei, J. W. Ward, K. P. Goetz, W. Li, M. Marcia, M. Guthold, J. E. Anthony, O. D. Jurchescu, P. J. Diemer, Z. A. Lamport, Y. Mei, J. W. Ward, K. P. Goetz, W. Li, M. M. Payne, M. Guthold and J. E. Anthony, 'Quantitative analysis of the density of trap states at the semiconductor-dielectric interface in organic field-effect transistors', *Appl. Phys. Lett.*, vol. 107, no. 10, p. 103303, 2015.

[143] D. M. Binkley, N. Verma, R. L. Crawford, E. J. Brandon and T. N. Jackson, 'Design of an auto-zeroed, differential, organic thin-film field-effect transistor amplifier for sensor applications', in *Opt. Sci. Technol. SPIE 49th Annu. Meet.*, Denver, CO, USA, 2004, pp. 41–52.

[144] S. Abdinia, M. Benwadih, E. Cantatore, I. Chartier, S. Jacob, L. Maddiona, G. Maiellaro, L. Mariucci, G. Palmisano, M. Rapisarda, F. Tramontana and A. H. M. van Roermund, 'Design of analog and digital building blocks in a fully printed complementary organic technology', in *Eur. Solid-State Circuits Conf.*, Bordeaux, France, 2012, pp. 145–148.

[145] T. Moy, L. Huang, W. Rieutort-Louis, S. Wagner, J. C. Sturm and N. Verma, '16.4 A flexible EEG acquisition and biomarker extraction system based on thin-film electronics', in *2016 IEEE Int. Solid-State Circuits Conf.*, San Francisco, CA, USA, 2016, pp. 294–295.

[146] C. C. Enz and G. C. Temes, 'Circuit techniques for reducing the effects of op-amp imperfections: Autozeroing, correlated double

sampling, and chopper stabilization', *Proc. IEEE*, vol. 84, no. 11, pp. 1584–1614, 1996.

[147] W. Xiong, U. Zschieschang, H. Klauk and B. Murmann, 'A 3V 6b successive-approximation ADC using complementary organic thin-film transistors on glass', in *2010 IEEE Int. Solid-State Circuits Conf.*, San Francisco, CA, USA 2010, pp. 134–135.

[148] S. Abdinia, M. Benwadih, R. Coppard, S. Jacob, G. Maiellaro, G. Palmisano, M. Rizzo, A. Scuderi, F. Tramontana, A. van Roermund and E. Cantatore, 'A 4b ADC manufactured in a fully-printed organic complementary technology including resistors', in *2013 IEEE Int. Solid-State Circuits Conf.*, San Francisco, CA, USA, 2013, pp. 106–107.

[149] T. Zaki, F. Ante, U. Zschieschang, J. Butschke, F. Letzkus, H. Richter, H. Klauk and J. N. Burghartz, 'A 3.3 v 6-bit 100 kS/s current-steering digital-to-analog converter using organic p-type thin-film transistors on glass', *IEEE J. Solid-State Circuits*, vol. 47, no. 1, pp. 292–300, 2012.

[150] S. Ouzounov, E. Roza, J. A. Hegt, G. van der Weide and A. H. M. van Roermund, 'Analysis and design of high-performance asynchronous sigma-delta modulators with a binary quantizer', *IEEE J. Solid-State Circuits*, vol. 41, no. 3, pp. 588–596, 2006.

[151] D. Raiteri, P. van Lieshout, A. van Roermund and E. Cantatore, 'An organic VCO-based ADC for quasi-static signals achieving 1LSB INL at 6b resolution', in *2013 IEEE Int. Solid-State Circuits Conf.*, San Francisco, CA, USA, 2013, pp. 108–109.

[152] M. Drack, I. Graz, T. Sekitani, T. Someya, M. Kaltenbrunner and S. Bauer, 'An imperceptible plastic electronic wrap', *Adv. Mater.*, vol. 27, no. 1, pp. 34–40, 2014.

[153] A. Perinot, P. Kshirsagar, M. A. Malvindi, P. P. Pompa, R. Fiammengo and M. Caironi, 'Direct-written polymer field-effect transistors operating at 20 MHz', *Sci. Rep.*, vol. 6, no. 1, p. 38941, Dec. 2016.

[154] N. Münzenrieder, L. Petti, C. Zysset, T. Kinkeldei, G. A. Salvatore and G. Tröster, 'Flexible self-aligned amorphous InGaZnO thin-film transistors with submicrometer channel length and a transit frequency of 135 MHz', *IEEE Trans. Electron Devices*, vol. 60, no. 9, pp. 2815–2820, Sep.2013.

Cambridge Elements ≡

Flexible and Large-Area Electronics

Ravinder Dahiya
University of Glasgow

Ravinder Dahiya is Professor of Electronic and Nanoscale Engineering, and an EPSRC Fellow, at the University of Glasgow. He is a Distinguished Lecturer of the IEEE Sensors Council, and serves on the Editorial Boards of the IEEE Sensors Journal and IEEE Transactions on Robotics. He is an expert in the field of flexible and bendable electronics, with a research focus on cost-effective approaches for high-performance electronics and sensors on flexible and bendable substrates.

Luigi G. Occhipinti
University of Cambridge

Luigi G. Occhipinti is National Outreach and Business Development Manager of the EPSRC Centre of Innovative Manufacturing for Large-Area Electronics, University of Cambridge. He is Founder and Director of Engineering at Cambridge Innovation Technologies Consulting Limited, providing innovation in high-tech fields that require multi-disciplinary approaches. He is a recognised expert in printed, organic, and large-area electronics and integrated smart systems with over 20 years' experience in the semiconductor industry, and is a former R&D Senior Group Manager and Programs Director at STMicroelectronics.

About the series

This innovative series provides authoritative coverage of the state of the art in bendable and large-area electronics. Specific Elements provide in-depth coverage of key technologies, materials and techniques for the design and manufacturing of flexible electronic circuits and systems, as well as cutting-edge insights into emerging real-world applications. This series is a dynamic reference resource for graduate students, researchers, and practitioners in electrical engineering, physics, chemistry and materials.

Cambridge Elements ≡
Flexible and Large-Area Electronics

Elements in the series

Bioresorbable Materials and Their Application in Electronics
Xian Huang
9781108406239

Organic and Amorphous-Metal-Oxide Flexible Analogue Electronics
Vincenzo Pecunia, Marco Fattori, Sahel Abdinia, Henning Sirringhaus, Eugenio Cantatore
9781108458191

A full series listing is available at www.cambridge.org/eflex

Printed in the United States
By Bookmasters